National Center for Construction Education and Research

Pipeline Maintenance Level Two

Prentice Hall
Upper Saddle River, NJ
Columbus, OH

contren™
Learning Series

American Petroleum Institute
Pipeliner Training and Assessment Program (PTAP)

This information is general in nature and intended for training purposes only. Actual performance of activities described in this manual requires compliance with all applicable operating, service, maintenance, and safety procedures under the direction of qualified personnel. References in this manual to patented or proprietary devices do not constitute a recommendation of their use.

Prentice Hall

nccer

contren™ Learning Series

10 9 8
ISBN 0-13-046678-6

Preface

This volume was developed by the National Center for Construction Education and Research (NCCER) in response to the training needs of the construction, maintenance, and pipeline industries. It is one of many in NCCER's *Contren™ Learning Series*. The program, covering training for close to 40 construction and maintenance areas, and including skills assessments, safety training, and management education, was developed over a period of years by industry and education specialists.

NCCER also maintains a National Registry that provides transcripts, certificates, and wallet cards to individuals who have successfully completed modules of NCCER's *Contren™ Learning Series*, when the training program is delivered by an NCCER Accredited Training Sponsor.

The NCCER is a not-for-profit 501(c)(3) education foundation established in 1995 by the world's largest and most progressive construction companies and national construction associations. It was founded to address the severe workforce shortage facing the industry and to develop a standardized training process and curricula. Today, NCCER is supported by hundreds of leading construction and maintenance companies, manufacturers, and national associations, including the following partnering organizations:

PARTNERING ASSOCIATIONS

- American Fire Sprinkler Association
- American Petroleum Institute
- American Society for Training & Development
- American Welding Society
- Associated Builders & Contractors, Inc.
- Association for Career and Technical Education
- Associated General Contractors of America
- Carolinas AGC, Inc.
- Carolinas Electrical Contractors Association
- Citizens Democracy Corps
- Construction Industry Institute
- Construction Users Roundtable
- Design-Build Institute of America
- Merit Contractors Association of Canada
- Metal Building Manufacturers Association
- National Association of Minority Contractors
- National Association of State Supervisors for Trade and Industrial Education
- National Association of Women in Construction
- National Insulation Association
- National Ready Mixed Concrete Association
- National Utility Contractors Association
- National Vocational Technical Honor Society
- North American Crane Bureau
- Painting & Decorating Contractors of America
- Portland Cement Association
- SkillsUSA-VICA
- Steel Erectors Association of America
- Texas Gulf Coast Chapter ABC
- U.S. Army Corps of Engineers
- University of Florida
- Women Construction Owners & Executives, USA

Some features of NCCER's *Contren™ Learning Series* are:

- An industry-proven record of success
- Curricula developed by the industry for the industry
- National standardization providing portability of learned job skills and educational credits
- Credentials for individuals through NCCER's National Registry
- Compliance with Apprenticeship, Training, Employer, and Labor Services (ATELS) requirements for related classroom training (CFR 29:29)
- Well-illustrated, up-to-date, and practical information

Acknowledgments

This module is a product of the farsightedness and leadership of the following sponsors:

Alyeska Pipeline Service Company
Bradford Brothers, Inc.
Ed Walton Construction Co.
Enron Transportation Services
Equistar Pipeline Operations
Green Equipment Company
Koch Pipeline Company, L. P.
Lakehead Pipeline Company
Marathon Ashland Pipe Line LLC
Matrix Service Mid-Continent
Plantation Pipe Line Company
Sun Pipeline
West Shore Pipe Line
Williams Energy Services

This curriculum would not exist were it not for the dedication and unselfish energy of those volunteers who served on the Authoring Team. A sincere thanks is extended to:

Ken Ballon
Lloyd Bell
Cecil Bradford, III
Robert Buchanan
Daniel Cerkoney
Ron Davenport
Vance Davis
Don Gilley
Greg Guy
Roy Harlan
Neal Jones
Bruce Larson
Grady McWhorter
Daniel Pallo
Philip Tideman
Thomas Trauthwein

A final note: This book is the result of a collaborative effort involving the production, editorial, and development staff at Prentice-Hall, Inc., and the National Center for Construction Education and Research. Thanks to all of the dedicated people involved in the many stages of this project.

Contents

Module 62201-02

Right-of-Way Inspection (CT 15 and 16)

Introduction to the Pipeline Operator Qualification Rule, API, and NCCER

It could be said that petroleum products keep America running. Petroleum products like oil and natural gas are the principal sources of energy in the United States and are used in many manufacturing and chemical processes. Petroleum products are primarily transported by pipeline. There are more than a million miles of liquid and natural gas pipeline in this country, ranging in diameter from two inches to four feet. These pipelines bring crude oil and natural gas from production fields to processing plants and refineries—and ultimately to the consumer.

Pipeline Operator Qualification

Petroleum products are considered hazardous, so pipeline personnel must take extra precautions and follow specific procedures when working with them. To help ensure the safe handling and transportation of petroleum products, the United States Department of Transportation (DOT) mandates safe practices as federal regulations.

One such regulation, the Operator Qualification (OQ) rule, requires pipeline operators to develop a written qualification plan to evaluate their personnel and their contract personnel (see *49 CFR Part 192 Subpart N—Gas,* and *49 CFR Part 195 Subpart G—Liquid*). The evaluation hinges on the operator's ability to perform certain covered tasks and to be able to recognize and respond to abnormal operating conditions that may be encountered while performing these covered tasks.

Covered Tasks

A pipeline performance task is deemed *covered* if the activity passes the following four-part test:

1. Is it performed on a pipeline facility?
2. Is it an operations or maintenance task?
3. Is it performed as a requirement of *49 CFR Part 195 (Liquid)* or *49 CFR Part 192 (Gas)*?
4. Does it affect the operation or integrity of the pipeline facility?

If the answer to all four questions is *yes,* then the activity is considered *covered* and must figure into the evaluation process of pipeline personnel.

American Petroleum Institute

The OQ rule and its related covered tasks provided the catalyst to develop a unified pipeline training and assessment program—so that pipeline operators and owners have ready access to the assessment and training tools they need. This effort was the direct result of the vision and leadership provided by The American Petroleum Institute (API) and API members.

Efforts to unify the industry are nothing new to API. Indeed, the impetus for forming the Institute in 1919 was the need to standardize engineering specifications for drilling and production equipment. Today, the most pressing issues revolve around public perceptions and government policies toward the industry. Speaking with one voice on these issues has become as essential as having interchangeable parts in the field. API's members determine the petroleum industry's positions on public policy as well as the standards that govern its day-to-day operations worldwide.

Consortium on Operator Qualification

API assembled a focus group whose primary purpose was to identify covered tasks and provide compliance recommendations to pipeline operators.

The Consortium on Operator Qualification (COOQ) became the guiding force for the identification and development of covered tasks for the liquid pipeline industry. Their work product, API publication 1161, *Guidance Document for the Qualification of Liquid Pipeline Personnel,* formed the basis for the resulting Pipeliner Training and Assessment Program.

The covered task list created in API publication 1161 has now evolved to include gas covered tasks and is maintained by NCCER. The most current covered task list can be downloaded at www.nccer.org.

National Center for Construction Education and Research

With the guidance and support of API, its members, and the Consortium on Operator Qualification, the National Center for Construction Education and Research (NCCER) was able to design and implement the Pipeliner Training and Assessment Program.

NCCER is a not-for-profit 501(c)(3) education foundation established in 1995 by the world's largest and most progressive construction and maintenance contractors and national contractor associations. NCCER was founded to address the

severe workforce shortage facing the construction and maintenance industries and to develop a standardized training process and curricula. Today, NCCER is supported by hundreds of leading construction and maintenance companies, manufacturers, and national associations and developed more than 35 construction and maintenance titles.

NCCER also maintains a National Registry, which contains the records of trainees and instructors who have received NCCER training. All completion information for every trainee is sent to the NCCER and filed within the Registry. The Registry can then confirm training and skills for workers as they move from company to company, state to state, or even within their own company.

Standardized Curriculum (NCCER's Contren™ Learning Series)

NCCER curricula are developed using the expertise of industry subject matter experts (SMEs). The use of SMEs as an integral part of every new development or revision ensures that the curriculum is current, industry-driven, and indicative of industry practices. The development and/or revision process is facilitated from start to finish by NCCER project managers, who work closely with the SMEs and a select team of technical writer(s) on the development and presentation of the curriculum.

Pipeline: Five Training Paths

The pipeline curriculum focuses on essential skills and competencies within five skill areas:

- Corrosion Control
- Electrical and Instrumentation
- Maintenance
- Mechanical
- Operations (Field and Control Center)

SMEs from the pipeline industry have determined the skills and competencies needed based on API's covered tasks and the consensus of the group/industry. Specific company information is generally not incorporated into the curriculum unless the SMEs determine that the information is the industry standard.

Training Modules and Covered Tasks

The smallest unit of instruction in the curriculum is a *module*. Each curriculum is composed of modules (the equivalent of stand-alone chapters) that are logically sequenced into complete *levels* of instruction. Every module is task-oriented and requires the successful completion of a written (knowledge-based) examination and a performance (hands-on) test. Modules may be individually selected and/or combined for efficient task training. Modules in the pipeline curricula, to the extent possible, will align with API-developed covered tasks. While not all modules will relate to covered tasks, all covered tasks will relate to at least one module.

Benefits of Training to the Pipeline Marketplace

Applying NCCER's standardized training process allows pipeline owners and contractors in the United States to establish the benchmark for acceptable skill and competency levels within the pipeline industry. This process will assist training organizations in developing and/or improving standardized pipeline training programs that effectively match their resources and operating conditions while responding to regulatory demands and to the needs of the industry. The process will also recognize training organizations that meet the minimum standards of competence, warranting regulatory, public, and professional confidence. This in turn will foster national unity within the pipeline industry for the mutual benefit of individual technicians, owners, contractors, and ultimately pipeline-users and society.

Additional Resources

To review the Pipeliner Training and Assessment Program Implementation and Resource Guide and the latest version of pipeline covered tasks, visit NCCER's Web site at www.nccer.org, or contact:

National Center for Construction Education and Research
PO Box 141104
Gainesville, FL 32614-1104
(888) 622-3720
Web site: www.nccer.org

For more information on API, please contact:

American Petroleum Institute
1220 L Street, Northwest
Washington, D.C. 20005
(202) 682-8000
Web site: www.api.org

To order *API Pub 1161*, contact Global Engineering Documents by calling 1-800-854-7149 or visiting their Web site at www.global.his.com.

Additional sources of information on the pipeline operator qualification regulation include:

DOT – Office of Pipeline Safety
www.ops.dot.gov

Transportation Safety Institute
www.tsi.dot.gov

COURSE MAP

This course map shows all of the modules in the second level of the Pipeline Maintenance curriculum. The suggested training order begins at the bottom and proceeds up. Skill levels increase as you advance on the course map. The local Training Program Sponsor may adjust the training order.

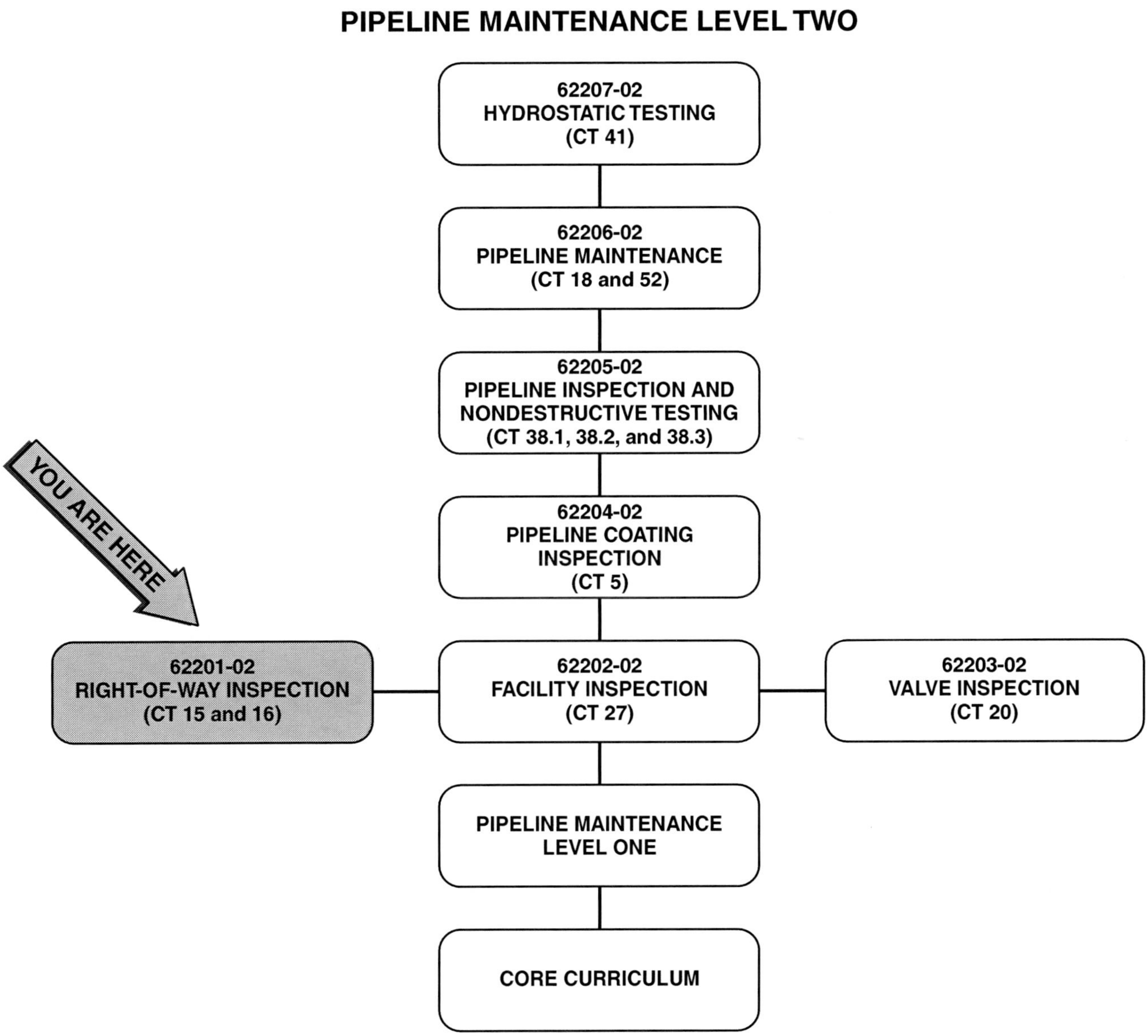

201CMAP.EPS

MODULE 62201-02 CONTENTS

Figures

Table

Right-of-Way Inspection (CT 15 and 16)

Objectives

When you have completed this module, you will be able to do the following:

1. Inspect navigable water crossings (CT 16).
2. Perform vehicle/foot and aerial patrols
3. Inspect surface conditions of the right-of-way (CT 15).

Prerequisites

Before you begin this module, it is recommended that you successfully complete the following: Core Curriculum; Pipeline Core, Modules 66101-02 and 66102-02; Pipeline Maintenance Level One.

Required Trainee Materials

1. Pencil and paper
2. Appropriate personal protective equipment

1.0.0 ◆ INTRODUCTION

The **pipeline right-of-way (ROW)** is the area along the pipeline that has been legally permitted to be used by the pipeline company to construct, operate, and maintain the pipeline. The ROW runs the entire length of the line and is wide enough to allow for construction equipment and cathodic protection equipment installed on the line and, in some cases, for access roads. Most ROWs vary greatly and can be from 10 to 100 feet wide. Always know what the actual distance of the ROW is before going out to work on it. After the pipeline is built, the ROW is returned to its original condition as much as possible, including restoring fences across it, so that the use of the adjoining land is not obstructed. The ROW itself is not fenced in, except for aboveground equipment and facilities. The ROW is patrolled regularly by plane, by vehicle, or on foot in order to observe conditions, construction activity, encroachments, markers, soil condition, uncovered pipe, facility condition and security, leaks, and any indications of problems.

Under *49 CFR Part 195.412 (a)*, each pipeline ROW must be inspected at least 26 times per year at intervals not exceeding 3 weeks. Under *49 CFR Part 192.705*, natural gas or other gas transmission line right-of-ways must be inspected periodically to ensure that they are being maintained in a safe manner. How often a right-of-way is inspected is determined by several factors, including the size of the line, the class location of the line, the terrain where the line is installed, and the weather conditions since the last inspection. All of these factors are taken into account when determining how often a right-of-way should be inspected, but the time frame between inspections should be no longer than the time intervals presented in *Table 1*.

Table 1 Maximum Intervals between ROW Inspections

Class Locations of Lines	At Highway and Railroad Crossings	All Other Places
1, 2	7½ months, but at least twice each calendar year	15 months, but at least once each calendar year
3	4½ months, but at least four times each calendar year	7½ months, but at least twice each calendar year
4	4½ months, but at least four times each calendar year	7½ months, but at least twice each calendar year

201T01.EPS

The following sections detail:

- Navigable water crossings inspections
- Methods of patrol
- Reportable observations
- Reporting protocol

2.0.0 ◆ NAVIGABLE WATER CROSSINGS INSPECTIONS

The Code of Federal Regulations requires that all navigable water crossings be inspected at least every five years. To ensure the integrity of the pipeline, water crossings may require inspections more frequently to determine the impact of flooding, upstream construction, or other activities. Inspections determine any changes in the horizontal alignment of the pipe and the river bottom profile, verifying that the pipe has adequate cover.

Navigable waterways are defined as those subject to commercial barge traffic. Inspections are performed by persons who are qualified to carry out such inspections and prepare permanent records of their findings. The inspections determine the pipeline location and any exposed pipe within the limits of the navigable waterway and the ROW immediately adjacent on each side. The permanent records must be of sufficient detail to compare with the previous crossing records to determine any change in the location of the pipeline. They must also detail the length and location of any exposed pipeline. The inspector prepares both plan and profile drawings detailing the findings of the inspection.

Navigable water crossing inspection includes pipeline location, surface inspection, and submarine inspection.

2.1.0 Locating the Pipeline

Pipeline is located by using existing drawings and probing/sonar equipment. Location of the pipeline may depend on the circumstances and may be accomplished with a combination of these methods.

2.1.1 Existing Drawings

Waterway inspections are performed with appropriate surveying equipment and techniques, and the results are compared with existing alignment drawings that show the crossings, the location of known landmarks, and the as-built river profile. The as-built river profile includes the pipeline location, the river bottom profile of previous surveys, the water levels of previous surveys, and the description and location of two bench marks on each side of the river as applicable. *Figure 1* shows an as-built river profile. The inspector reestablishs the baseline indicated on the plan drawings and accurately determines the route of the pipeline through the river with reference to this line. The length of this traverse extends between convenient points on the banks that are above the water level.

The river bottom and banks between reference points are profiled over the pipe with elevations referenced to the existing bench marks shown on the drawings. If an existing bench mark has been destroyed, a new bench mark is set in as permanent a position as possible. Profile data also includes the elevation of the water level and the top of pipe at all exposed and suspended sections. Vertical measurements for the profile may be made with a portable fathometer or weighted sounding chain. Location of the pipeline may be assisted by visual examination of adjoining banks to search for exposed pipe, crossing signs, and bench marks.

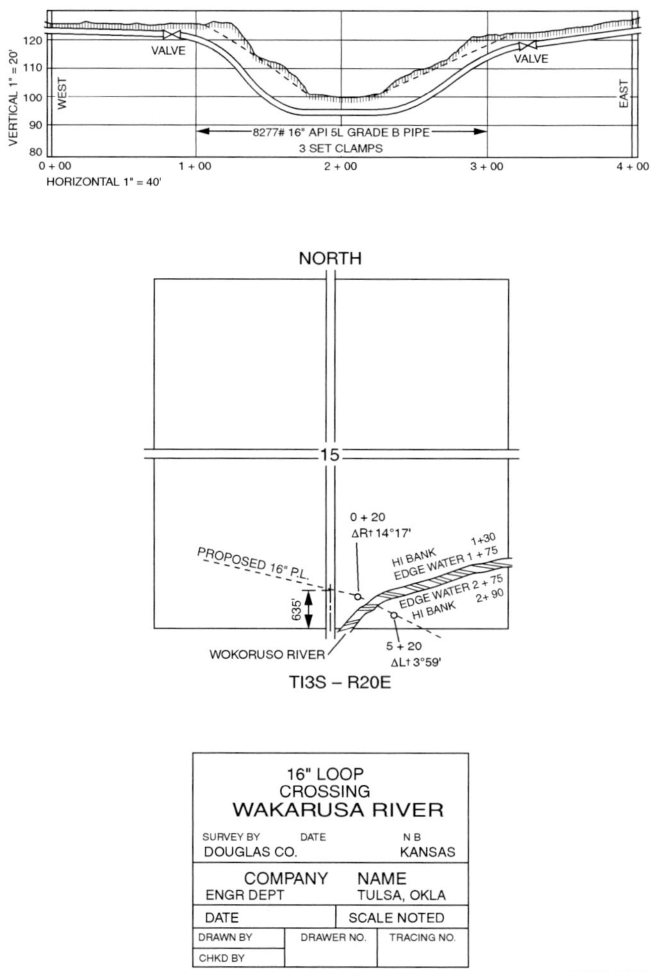

201F01.EPS

Figure 1 ◆ As-built river profile.

NOTE
Ensure that all maps, survey records, and drawings are up-to-date.

2.1.2 Probing/Sonar Equipment

Probing equipment used to locate the pipeline may include mechanical, induced-tone, or transmitter/receiver types. Probing equipment, such as a probe rod (*Figure 2*), is used to physically probe the submerged pipeline to determine the depth of cover and to verify the location of the pipeline. Damage to pipeline coating can be prevented with proper training in probe use. When using probing equipment:

Step 1 Locate the landfall of the pipeline crossing to establish a reference point.

Step 2 Probe the submerged pipeline to determine the amount of cover and to verify the location of the pipeline. Probe for the location of the river bottom and the top of the pipe.

Step 3 Document the findings, including:

- Location of exposed or unsupported pipe
- Depth of cover
- Pipeline damage
- Debris hung up on the pipeline

CAUTION
Make the proper company notifications upon the discovery of unsupported submerged pipe.

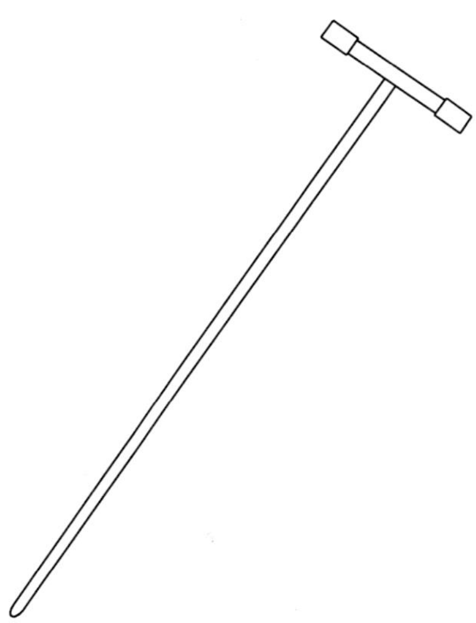

Figure 2 ◆ Probe rod.

Specialized sonar equipment can also be used to determine pipeline location and cover by transmitting a sound signal and monitoring the received signal that returns. When using sonar equipment:

Step 1 Locate the landfall of the pipeline crossing to establish a reference point. Ensure that permanent horizontal and vertical control monuments are established on both banks. Accurately measure and record the pipeline horizontal and vertical alignment, depths of cover, and waterway profile and alignment.

Step 2 Take sonar readings of the pipeline to determine the amount of cover and to verify the location of the pipeline.

Step 3 Document the findings for comparison to previous inspections. Detail the condition of the crossing, any remedial action taken, and problem areas that may need additional consideration.

Step 4 Determine if additional underwater diver inspections are required.

2.2.0 Adjacent Bank Inspection

Adjacent bank inspection of waterways includes inspection of adjacent banks and ROW. Surface conditions to be reported include:

- Evidence of release of product
- Evidence of excess vegetation
- Unusual activities
- Evidence of erosion
- Exposed pipe segments
- Missing or damaged signs
- Missing or damaged supports for aboveground pipe
- Other unsafe conditions

2.3.0 Submarine Inspection

During submarine inspection, a diver passes over each pipe at intervals of 20 feet or less, and if exposed or suspended pipe is located, the diver follows these sections with continuous passes. The size and location of any build-up of debris against exposed or suspended pipe section, the condition of pipe and/or pipe coating, the nature of the river bottom at the pipe, and any other underwater condition that may appear to adversely affect the safety of the pipeline crossing are also determined and reported. Two such underwater conditions are scouring of the river bottom and undercutting of the banks. Sonar equipment, instead of a diver, may be used to detect exposed pipe.

A device used in undersea inspection is the **submersible**. This is essentially a two-man mini-submarine. The chief advantage of using a submersible instead of divers is it allows personnel to dive deeper and for longer periods. The submersibles can also carry photographic and sonar equipment.

2.4.0 Navigable Waterway Spans Inspection

Navigable waterway overhead spans must be inspected for proper alignment and attachment points from bank to bank. Always check all supports and cables used to secure and position the span. *Figure 3* shows a navigable waterway span.

3.0.0 ◆ METHODS OF PATROL

The pipeline ROW is patrolled by plane, by vehicle, or on foot. The frequency of aerial, vehicle, and foot patrols for liquid lines must satisfy the requirements of *49 CFR Part 195.412 (a)*. The sum of all vehicle, foot, and aerial patrols must be at least 26 times per year and at least every 3 weeks for each pipeline ROW.

As per *49 CFR Part 192.705*, the frequency of patrols for gas pipelines is determined by the size of the line, operating pressures, class location, terrain, weather, and other relevant factors.

Some conditions, such as floods, significant increases in third-party construction or agricultural activities, or evidence or threats of vandalism may warrant scheduling more frequent inspections. Some inspections occur more often according to the product and population density. The results of all patrols are reported and documented.

3.1.0 Aerial Patrol

The most common form of information gathering is done by aerial patrol on a weekly or biweekly schedule according to DOT and company policy. During these patrols, the inspectors watch for and report developments or changes on and adjacent to the ROW that may result in the need

201F03.EPS

Figure 3 ◆ Navigable waterway overhead crossing.

for maintenance or investigation. Often, the first sign of an underground leak is an area of wilted vegetation. Soil erosion, stream bed changes, weathering of markers and signs, and growth of bushes and trees are gradual changes that will eventually require maintenance. Construction work near the ROW is noted and reported. Local pipeline personnel investigate such activities and make arrangements to prevent pipeline damage.

Finding leaks, reporting their locations, and estimating their size and nature are the prime responsibilities of air patrol inspectors. Special flights are made to investigate potential leaks. Communication between the aircraft and repair crew headquarters makes fast action possible in emergencies.

If a threatening activity is observed, the inspector communicates a warning and/or instructions to the persons involved and then contacts the control room. Any threatening conditions observed are reported immediately according to company policy.

3.1.1 Qualifications for Aerial Patrol

Qualifications for aerial patrol vary for different companies and different areas. However, some qualifications are required based on the aircraft type, inspector qualifications, and flight frequency. Aircraft of a type and quality suitable for the intended purposes and local geographic needs are used to patrol the pipeline.

The inspector must be trained to recognize conditions to be reported and to follow the reporting procedures of their company. They must develop good communications with company operations staff to obtain an effective working relationship. Discussions with staff before each flight about possible problems or areas of interest can increase the effectiveness of aerial patrols.

3.1.2 Methodology

Because of the variety of geographic and physical circumstances, details of speed, altitude, weather, time of day, direction of flights, flight schedules, and other factors must be determined by the inspector and local requirements.

Photographs of observations are encouraged where practical and should be used to document any leak or discharge, damage to the pipeline or ROW, or any special situations.

Inspector and company staff communications are the primary method of assessing the effectiveness of aerial inspection. It is sometimes necessary to have staff accompany the inspector to observe and validate observations. Before each flight patrol, the inspector must communicate with the operations staff responsible for the subject areas in order to discuss any possible problems or areas of special interest.

The types of equipment normally required for aerial patrols include relevant pipeline maps, a cell phone, a pager, and a camera. Global positioning system (GPS) equipment is becoming more frequently used.

3.2.0 Vehicle/Foot Patrol

Linewalker is the name given to the ground patrol who may walk or use a vehicle to inspect the line. Although aerial patrol has taken over much of line inspection, linewalkers are still needed in certain areas, such as densely populated areas where flying at a low altitude may not be allowed. Covering the ROW at ground level may be required when emergencies occur during darkness or when bad weather prevents aerial patrol.

Vehicle/foot patrol duties include observing the ROW conditions, construction activity at or near the line, encroachments, line marker conditions, soil erosion, uncovered pipe, pipeline facility condition, leaks, and any other condition that would place the pipeline or public in danger. Faulty security, such as suspicious activities or damaged fencing, gates, and locks, should also be checked.

Vehicle and foot patrols are a good way to check the conditions closely. Reliable and useful information about specific situations is often obtained from on-site surveys by experienced maintenance crews. They observe conditions and make notes of conditions on every trip to or along the pipeline ROW.

4.0.0 ◆ REPORTABLE OBSERVATIONS

The inspector checks the surface conditions on or adjacent to each pipeline ROW and reports appropriate observations. In cases other than emergencies, the report is written from the inspector's log. The notable conditions to be reported on patrols may include:

- **Emergency situations**/evidence of release
- Environmental changes
- Damaged or exposed pipeline
- Missing or damaged signs/markers
- **Encroachment**/nearby activities
- Vandalism
- Other unsafe conditions

4.1.0 Emergency Situations/ Evidence of Release

The definition of an emergency situation is a leak, evidence of a possible leak, or activities that pose an immediate threat to the pipeline or to public safety. A possible leak observation might be petroleum on the ground, vapors in the air, wilted or dead vegetation over the pipeline, dead or sick animals near the pipeline, the smell of hydrocarbon, or petroleum sheen on water. If the leak stems from buried pipe, it usually takes some time for the evidence to show on the surface. Any evidence of a leak requires an immediate verbal report, as well as written reports. Company policy will dictate response and repair.

WARNING!
If an unexpected hazardous liquid or carbon dioxide is encountered, local emergency personnel should be notified immediately if appropriate. Additionally, the proper company notifications should be made, and the leak should be identified if possible.

4.2.0 Environmental Changes

Environmental changes may be expected due to local conditions. These types of changes include:

- Washouts over or along the pipeline
- Riverbank erosion
- Excessive vegetation along the right-of-way
- Creek or riverbed location changes
- Earth movement or subsidence
- Trees and debris clogging waterways

4.2.1 Evidence of Erosion

Soil erosion caused by water and wind and washouts caused by flood waters may expose or threaten to expose the pipeline. Preventive measures may need to be taken to control the erosion. These measures include terraces for surface erosion, fences for drifting sand, and bulkheads for stream and river bank erosion. Evidence of erosion must be reported so that these measures can be put in place before the pipe is exposed.

4.2.2 Evidence of Excess Vegetation

Most ROW policies limit or ban trees growing on the ROW. Any vegetation that grows excessively near or on the ROW may hamper inspection or threaten to block access for maintenance and repair. This excessive vegetation growth must be reported per company policy.

4.3.0 Pipeline Damage or Exposure

Pipeline damage may be discovered directly on aboveground pipe. Evidence of a leak is indicative of damaged underground pipe. Any oil, petroleum product, vapor, or other evidence of spills or discharges from the pipeline, main-line block valves, interface detectors, and other facilities must be reported at once. Other evidence of leaks may include:

- Oil or product on a water surface adjacent to the pipeline
- Waste oil, salt ponds, or drainage
- Downed livestock
- Wilted or discolored vegetation
- Vapor cloud

During an aerial patrol, some damage may not be readily observed. The inspector must be alert to subtle changes that warrant reporting for closer inspection.

Pipeline exposure may be caused by environmental conditions, such as weather or earth movements. Exposed pipe should be reported for inspection and reburial. Whenever exposed pipe is discovered, it must be inspected for evidence of external damage, corrosion, and the extent of the damage or corrosion. Corrective action is taken according to your company policy. *Figure 4* shows exposed pipe.

4.3.1 Damaged or Missing Supports

Supports are required on various aboveground pipeline structures to provide protection from the forces exerted by anticipated loads. Supports may be found at overhead crossings, spans over ditches, scraper traps, block valves, manifold

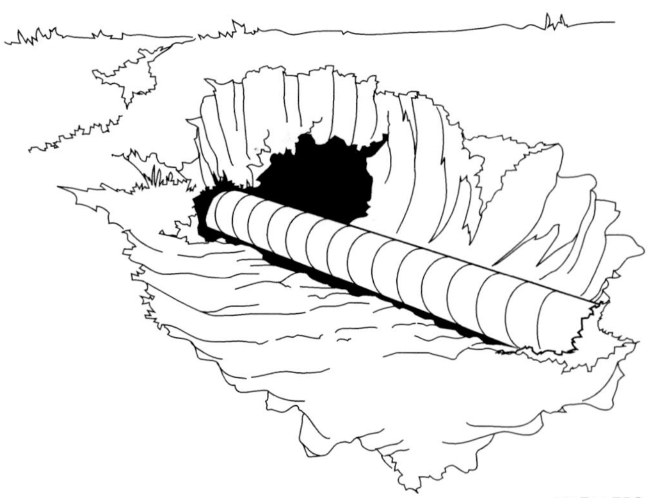

Figure 4 ◆ Exposed pipe.

valves and piping, and other structures. To be effective, these supports must be undamaged and in the correct position. Damaged or missing supports must be reported and repaired according to company policy. Some support problem situations may be emergencies if there is a threat of imminent collapse. These situations may require temporary supports to be installed and/or evacuation of the area. The cause of the damage should be investigated; it may have occurred naturally, resulted from vandalism, or resulted from third-party activity. *Figure 5* shows improper and misaligned supports.

The pipeline structure that should be supported must also be examined for damage. There may be evidence of unusual stress on the pipe or structure, or it may have been damaged by the same cause as the support. Some evidence to look for on the pipe or structure includes wrinkling, buckling, dents, and gouges. These types of damage must also be reported and repaired.

4.4.0 Missing or Damaged Signs/Markers

Signs and/or markers are required at all aboveground pipeline facilities, at all road, rail, and water crossings, and at sufficient intervals along the ROW to identify the pipeline location. They should, when possible, be directly over the pipeline, securely installed, and clearly legible. The information on the signs or markers must be up-to-date and correct. Aerial markers may also be present for use by aerial patrols. Marker locations are generally indicated on company maps.

Signs may become damaged by weather or other natural conditions, or they may be vandalized or stolen. During inspection patrols, any damaged, missing, improperly located, or inaccurate sign must be reported and replaced according to company policy.

4.5.0 Encroachment/Nearby Activities

The original activities along the pipeline are taken into account when starting pipeline operation, and most of the ROW is restored to its original usage after pipeline construction.

Encroachment is activity on the company's ROW carried out by outside parties. A particular encroachment might be authorized or not, so each situation requires specific handling. Unauthorized activities on the pipeline ROW that may affect the pipeline are often spotted by patrols doing line inspection and are reported to management. The unauthorized encroaching party is advised to cease activity until consent can be given by the

IMPROPER SUPPORTS

MISALIGNED SUPPORT

MISALIGNED SUPPORT

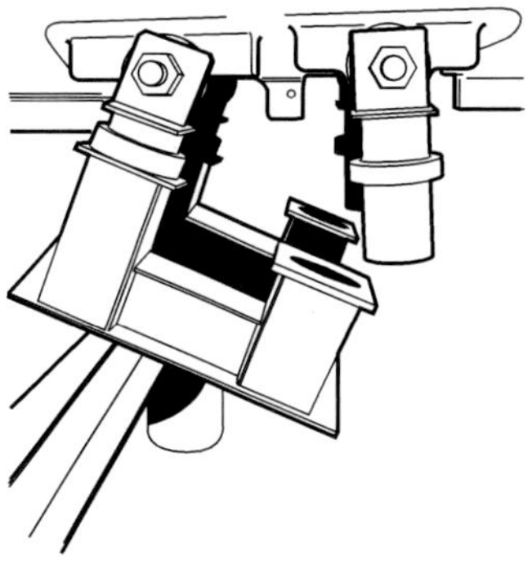

MISALIGNED SUPPORT

201F05.EPS

Figure 5 ◆ Improper and misaligned supports.

company. The company then investigates the activity and communicates with the encroaching party about a final resolution. If an immediate danger is posed by such activities, they must be stopped, and an emergency call must be made to the control center. Foreign crossings are reviewed and approved. Specific company procedures will apply to the crossing to ensure that the company pipeline is not damaged or compromised.

Reportable activities include any that might cause hazards or restrictions to the safety, integrity, and operation of the pipelines and easements. Besides encroachment, other activities that may threaten the pipeline security include:

- Blasting or surface mining within one-quarter mile of the pipeline
- Excavations within the ROW or excavations threatening the pipeline
- Clearing of timberland over, across, or along pipelines
- Building of terraces, water ponds, or pits within an area defined by company policy
- Construction of buildings, well derricks, residences, fences, or other structures within an area defined by company policy
- Brush or tree clearing, excavation, or ditching for new roads, utilities, or structures
- Grading, construction, or resurfacing of roads
- River dredging or water craft anchored over river crossings
- Fires occurring within an area defined by company policy
- Agricultural deep-ripping or subsoil plowing

4.6.0 Vandalism

Evidence of activity around a block valve or any other place along the pipeline ROW might indicate a line tap, vandalism, or sabotage. All surface facilities, such as pump stations and breakout tanks, should be provided with protection against vandalism or other unauthorized entry. Fencing may be present and should be checked by patrols for damage. Signs are required around these exposed facilities and should also be checked.

4.7.0 Other Unsafe Conditions

Other unsafe conditions that may be encountered along the ROW must also be reported. These include, but are not limited to, localized or general corrosion, pipe movement caused by environmental causes, material defects or physical damage, equipment malfunction, and events impacting the cover. The last condition includes airplane crashes, train wrecks, and heavy equipment traffic.

5.0.0 ◆ REPORTING PROTOCOL

The inspection records provided by the inspector are retained for at least five years or according to company policy. The results are compared with existing records to determine if any changes have occurred and if such changes require repairs. It is important and required that records of ROW inspections be made and maintained. The reporting protocol includes written reports, emergency and verbal reports, ROW investigation reports, inspector reports, drawings, and recordkeeping.

5.1.0 Written Reports

Patrol inspectors should make a written report of all specified and other material observations as soon as practical following the inspection. The report is then distributed to all relevant offices. The report should contain the following information:

- Date
- Inspector's name
- Identification of pipeline(s), section, or segment
- Location of observation
- Time of observation
- Remarks and description
- Report number
- Any actions taken
- Identification of any verbal reports

The inspection and any actions are recorded on a pipeline inspection report form, a form that asks for information on the reason for inspection, pipe and material conditions, general corrosion, defects or damage, and defect, damage, or corrosion measurements. A sketch of the damaged pipe is also required. The form is dated and signed by the inspector. *Figure 6* shows a pipeline inspection report.

5.1.1 Navigable Waterway Crossings

For navigable waterway crossings, reports of the survey and inspection are forwarded in individual written reports for each crossing site. Reports are submitted in triplicate to management and contain:

- A list of all exposed and suspended pipe sections, located by station numbers and showing clearance between the pipe and river bottom
- Identification of changes from the last inspection
- Plan and profile sheets of each pipeline surveyed at the crossing site, one being used for each single pipeline

PIPELINE INSPECTION REPORT

Frequency: At Time of Inspection

PART D — DEFECT, DAMAGE OR CORROSION INVESTIGATION

Notes:
1) Tabulate all Dents, Gouges, Slivers, Corrosion, S.C.C., etc.
2) List Reference Point - Valve, Tee, Center Road etc., as shown on Alignment Sheet with appropriate Station Number, if available.
3) Indicate Location of Longitudinal Seam. If Pipe is Vertical Substitute Compass direction for o'clock Position. For Spiral Weld Pipe, Note as Such.
4) Defects Such as Dents, Slivers, Gouges, etc., shall be tabulated individually. Corrosion Cracks and/or Corrosion Pits Closely Grouped May be Tabulated as One Defect.
5) Defect Width and Length Will be Measured Along the Girth and Longitudinal Axis Respectively.

Reference Point: ☐ Upstream End (Station No.) ____________ Description ______________________

☐ Downstream End (Station No) ____________ Description ______________________

Upstream End | Downstream End

SKETCH DEPICTING PIPE CUT AT BOTTOM AND LAID FLAT
NOTE: Indicate Girth Weld Location If Known

O'Clock Position →

← Length of Inspected Area →

PART E — DEFECT, DAMAGE OR CORROSION MEASUREMENTS AND LOCATION

No.	Type	Distance from Girth Weld	Distance from Long Seam	Width (Circumferencial)	Length (Longitudinal)	Maximum Depth	Number of Anomalies
1.							
2.							
3.							
4.							
5.							
6.							
7.							
8.							
9.							
10.							
11.							
12.							
13.							
14.							

If Repair is Required, send to Division Manager for Approval.

Repair Required ☐ No ☐ Yes Type of Repair ______________________

Approved by ______________________ Title ______________________

Remarks: ______________________

201F06.EPS

Figure 6 ◆ Pipeline inspection report (1 of 2).

PIPELINE INSPECTION REPORT

Frequency: At Time of Inspection

PART A — LOCATION

Reason For Inspection ____________

1. Location: Division ____________ County ____________ State ______ Date of Inspection ____________

A) **Mainline**
Alignment Sheet ________ Pipeline Size/Segment ________
Beg. M.P. ________ End M.P. ________
Beg. Sta. No. ________ End Sta. No. ________

Note: Station numbers shall be measured from a known landmark (Road Crossing, etc.)

B) **Company Facility: Pump Station, Tank Farm, etc.**
Location Name ____________
Physical Location of Pipe Inspected ____________
Milepost ____________

PART B — MATERIAL/CONDITIONS

2. Piping:
Pipe Size ________ In. O.D. Nom. W.T. ________ In. Actual W.T. ________ Grade ________

3. External Pipe Coating:

A) Coating
☐ Bare
☐ LAC 450
☐ Taped
☐ Insulated
☐ Other ________

B) Application
☐ Mill Coated
☐ Yard Coated
☐ Field Coated
☐ Other

C) Coating Thickness
Factory Applied____ mile
Field Applied ____mile

D) Condition of Coating
☐ Good ☐ Fair
☐ Poor

E) Bond
☐ Good
☐ Fair
☐ Poor

F) Cause of Coating Failure
☐ Damage
☐ Disbondment
☐ Defective Application
☐ Decomposition
☐ Other ________

4. Environment
☐ Clay ☐ Loam ☐ Sand ☐ Gumbo ☐ Caliche ☐ Rock ☐ Other ________
☐ Wet ☐ Dry ☐ Moist Depth of Cover (Inches) ________
Soil Resistivity Measurement ________ (ohm-cm) Measurement Method ________

5. Coating Repaired ☐ No ☐ Yes
Repair System ____________

6. Pipe-to-Soil Potential at Inspected Location (At Grade) ________ (volts) (At Pipe/Soil Interface) ________ (volts)

7. Were any Non-Destructive Tests Completed? ☐ No ☐ Yes Type ________

8. Evidence of: ☐ Corrosion ☐ Defect ☐ Damage ☐ Stress Corrosion Cracking (S.C.C.) ☐ Calcareous Deposits

9. Test Station Added ☐ No ☐ Yes I.D. No. ________

NOTES:
1) If S.C.C. or Corrosion is Noted, Complete Below and Back of Report (External Only, See Note 4)
2) If Defect or Damage is Noted, Complete Back of Report
3) If Repairs are Required, AAPL Form 502 "REPAIR REPORT FOR LEAK DAMAGE, CORROSION OR TEST FAILURE" Must be Attached.
4) AAPL Form 508 "INTERNAL CORROSION INSPECTION REPORT" Should be Completed When Internal Corrosion Is Detected.

PART C — EXTERNAL CORROSION

10. General Corrosion Information:

A. DESCRIPTION
☐ Less than 10% of Norm. Wall Thickness, Complete Front of Report
☐ Greater than 10% of Norm. W.T., Complete Front and Back of Report
☐ Pitting ☐ General ☐ Localized
Max Depth ________ (In.) Longitudinal Length ________ (In.)

B. STATUS
☐ Active
☐ Inactive
☐ Unknown

C. CAUSE
☐ Galvanic
☐ Stray Current
☐ Unknown
☐ Other ________

11. Cathodic Protection: ☐ Impressed ☐ Galvanic

12. Pipe-to-Soil Potential:
a) Last Pipe-to-Soil Measurement at Nearest Point Each Side of Corroded Area
Upstream ________ (volts) Downstream ________ (volts)
b) Distance from Corroded Area to Measurement Points:
Upstream ________ (feet) Downstream ________ (feet)
c) Date of Measurements: Upstream ________ Downstream ________

13. pH of Soil Near Corroded Area ________ pH of Moisture Under Coating ________

14. Pipe Surface Temperature if Fluid is Heated ________ (deg F)

Remarks ____________

Distribution: (Original) Corr. Ctrl./Eng (Copy) Division Office (Copy) Other ________

________ *Technician or Inspector* __/__/__ *Date* ________ *Supervisor* __/__/__ *Date*

SAMPLE

201F06.EPS

Figure 6 ◆ Pipeline inspection report (2 of 2).

5.2.0 Emergency and Verbal Reports

An emergency situation, such as a possible leak, requires the inspector to make an immediate cell telephone or radio telephone call to the appropriate contact. If the situation involves excavation or other activities performed by construction workers on the ground, the inspector should first warn the workers. After the first call, a second call is made to the company personnel designated to receive such calls. The verbal report is followed up with a written report.

Possible leak observations include petroleum product, vapor, or other discharge; dead vegetation over the pipeline; dead livestock or other dead animals near the pipeline; or petroleum sheen on nearby water. Emergency observations include activities or circumstances, such as construction activity, that may pose a threat to the integrity of the pipeline or to public safety.

Emergency verbal reports should include:

- Inspector's name
- Line section name and flight identification
- Location of the observation referenced to landmarks, geography, or city
- GPS location
- Description of the observation
- Indication that the observation is deemed an emergency

WARNING!

In case of a fire or explosion caused by the ignition of hydrocarbons, local emergency personnel should be notified immediately. Access to the area should be limited as much as possible, and the proper company notifications should be made.

5.3.0 ROW Investigation Reports

Personnel are dispatched to investigate the reported conditions and to fill out investigation reports, which are signed by these personnel to verify that conditions have been investigated. This is called an ROW investigation report and is filed according to DOT guidelines. A hard copy of the report is given to the originating inspector to check against the log. The original ROW investigation report goes to the division compliance records coordinator and then to division DOT files. The report details the reported information, the date and result of the investigation, and the action taken. The form is signed and dated by the inspector. *Figure 7* shows an ROW investigation report.

5.4.0 Inspector Reports

The inspector's report is distributed to all relevant offices. The report should contain the following information:

- Date
- Inspector's name
- Identification of pipeline(s), section, or segment
- Location of observation
- Time of observation
- Remarks and description
- Report number
- Identification of any verbal reports

The report number is assigned by the inspector, who will also indicate if the sighting is a new one or an old one. *Figure 8* shows a pipeline aerial patrol report.

5.5.0 Drawings

Drawings show unusual conditions, such as exposed pipe, and the location and elevation of bench marks used by the inspector. Drawings of navigable water crossings include the river bottom and adjacent banks between reference points and all elevations from previous inspections, where known.

5.6.0 Recordkeeping

The company staff should take timely and appropriate action in response to patrol observations and maintain a record of the company response. Such disposition records reference or are linked to the original reports and indicate what action was or will be taken and the status or disposition of the item observed. Records of the original observation, investigation reports, and the company action taken are maintained according to DOT and company policy and for at least two years.

RIGHT-OF-WAY INVESTIGATION REPORT

Frequency: As Required by Aerial Patrol Report

Report Date ____________

Note: Right-of-Way personnel should be consulted before confronting enroachment situations for advice pertaining to right-of-way restrictions and agreements.

PART A — REPORTED INFORMATION

1. Right-of-Way Investigation Report Number: ____________
2. Date Condition Reported: ____________
3. Milepost Location Reported: ____________
4. Descriptive Location Reported: ____________
5. Problem Reported: ____________

PART B — INVESTIGATION INFORMATION

1. Date Condition Investigated: ____________
2. Actual Milepost Location: ____________
3. Actual Location of Condition: ____________
4. Actual Problem Found: ____________

☐ Encroachment Advisory (AAPL Form 202) was given to encroaching party

Name of Encroaching Party: ____________

PART C — ACTION TAKEN

____

PART D — DISTRIBUTION INFORMATION

(Original)	Division Compliance Records Coordinator
(Copy)	Forward to Pilot at the Arizona Division Office
(Copy)	Right-of-Way

Technician or Inspector ____________ *Date* ____________ *Supervisor* ____________ *Date* ____________

Retention: Two Years

201F07.EPS

Figure 7 ◆ ROW investigation report.

ALL AMERICAN PIPELINE COMPANY
AAPL Form 204
Rev. 12/97

PIPELINE AERIAL PATROL REPORT

Sample of information format distributed through the computer network

Date: ______________________

LOCATION	TIME	OLD / NEW	REMARKS	REPORT NO.
CADIZ	05:53		STATION OK. ROW OK.	
LUDLOW	06:15		STATION OK. 1) ½ MILE EAST OF MP 145, DIGGING 50' NORTH OF LINE, TOOK PICTURE, REPORTED TO BARSTOW.	060391-1
12 GAUGE	06:49		STATION OK, ROW OK.	060391-2
MOJAVE	07:10		STATION OK. 1) ½ MILE EAST MP 36, NEW FENCE CROSSING ROW. REPORTED TO UNIT #205.	
TEJON	07:32		STATION OK. ROW OK.	
EMIDIO	07:43		STATION OK. ROW OK.	
PENTLAND	07:50		STATION OK. ROW OK.	
16"			1) DIGGING IN 3 PLACES NEAR LINE BETWEEN MP 38 & MP 39. REPORTED TO UNIT #237.	060391-3
BELRIDGE	08:11		STATION OK. ROW OK. DEADHEAD TO 30" LINE.	
MP 49	09:02		FOG. DEADHEAD TO LA PAZ.	
LA PAZ	13:16		STATION OK 1) MP 379, NUMBER ON GROUND.	
GILA	14:08		STATION OK. ROW OK.	
COOLIDGE	14:44		STATION OK. ROW OK.	
TOM MIX	14:57		STATION OK. ROW OK. END PATROL.	
DATE: 06/05/98				
TOM MIX	05:15		STATION OK. ROW OK.	
HOT SPRINGS	05:43		STATION OK. ROW OK.	
LORDSBURG	06:30		STATION OK. ROW OK.	
ANTHONY	07:24		STATION OK. ROW OK.	

201F08.EPS

Figure 8 ◆ Pipeline aerial patrol report.

Summary

The area along the pipeline that can be legally used by the pipeline company to construct, operate, and maintain the pipeline is called the right-of-way. Pipeline ROWs must be inspected, and these inspections may be performed during aerial, vehicle, and foot patrols.

Navigable water crossings must be inspected at least every five years. To ensure the integrity of the pipeline, water crossings may require more frequent inspections to determine the impact of flooding, upstream construction, or other activities. Inspections determine the depth of cover, the horizontal alignment of the pipe, and the river bottom profile.

Aerial, vehicle, and foot patrols of liquid pipelines must be performed 26 times a year at intervals not exceeding 3 weeks. The inspection frequency of liquid pipelines is determined by factors such as the size of the line, operating pressures, class location, terrain, and weather.

When patrolling a pipeline ROW, the inspector must check for certain conditions. These conditions include emergency situations/evidence of release, environmental changes, damaged or exposed pipeline, missing or damaged signs/markers, encroachment/nearby activities, and vandalism.

It is required that records of ROW inspections be made and maintained. Written reports of each inspection should be submitted per company policy. These reports are compared with existing records to determine if any changes have occurred and if such changes warrant repairs.

Review Questions

1. The area around the pipeline that has been legally permitted to be used by the pipeline company is the _____.

a. bypass
b. right-of-way
c. pipe section
d. clearing

2.. Navigable water crossings must be inspected at least every _____ years.

a. five
b. six
c. eight
d. ten

3. Probing equipment is used to locate a pipeline and to determine the _____.

a. coating type
b. pipe materials
c. product
d. depth of cover

4. Liquid pipelines must be inspected at least _____ times each year.

a. 10
b. 12
c. 24
d. 26

5. During submarine inspection, a diver passes over each pipe at intervals of _____ feet or less.

a. 20
b. 25
c. 10
d. 15

6. Liquid pipeline inspections should occur at least every _____ weeks.

a. two
b. three
c. four
d. five

7. _____ patrol is the most common manner of gathering information about the pipeline ROW.

a. Vehicle
b. Aerial
c. Navigable waterway
d. Foot

8. The ground patrol that walks or drives the line is a(n) _____.

a. inspector
b. technician
c. linewalker
d. line inspector

9. Damaged or missing pipeline supports must be reported and repaired according to _____.

a. DOT policy
b. state regulations
c. federal regulations
d. company policy

10. The proper locations for all pipeline markers are generally shown on _____.

a. aerial photos
b. pipeline diagrams
c. company maps
d. signs

11. _____ is activity on the company's ROW carried out by outside parties.

a. Encroachment
b. Trespass
c. Entrenchment
d. Vandalism

12. _____ copy/copies of a navigable waterway crossing report are submitted to management.

a. Four
b. Three
c. Two
d. One

13. If an emergency situation involves activities performed by workers on the ground, the inspector should first _____.

a. warn the workers
b. call local emergency personnel
c. make the proper company notifications
d. submit a verbal report

14. ROW investigation reports are filed according to _____ guidelines.

a. federal
b. state
c. DOT
d. OSHA

15. Records of observations and investigation reports are maintained for at least _____.

a. 6 months
b. 1 year
c. 18 months
d. 2 years

GLOSSARY

Trade Terms Introduced in This Module

Emergency situation: A leak, evidence of a possible leak, or activities that pose an immediate threat to the pipeline or to public safety.

Encroachment: To advance beyond proper or former limits. The term used to describe unauthorized activities being carried out on the pipeline right-of-way.

Linewalker: The ground patrol who walks or uses a vehicle to inspect the line.

Navigable waterways: Waterways that are subject to commercial waterway traffic.

Right-of-way (ROW): The area along the pipeline that has been legally permitted to be used by the pipeline company to construct, operate, and maintain the pipeline.

Submersible: A two-man mini-submarine.

CONTREN™ LEARNING SERIES — USER UPDATES

The NCCER makes every effort to keep these textbooks up-to-date and free of technical errors. We appreciate your help in this process. If you have an idea for improving this textbook, or if you find an error, a typographical mistake, or an inaccuracy in NCCER's Contren™ textbooks, please write us, using this form or a photocopy. Be sure to include the exact module number, page number, a detailed description, and the correction, if applicable. Your input will be brought to the attention of the Technical Review Committee. Thank you for your assistance.

Instructors – If you found that additional materials were necessary in order to teach this module effectively, please let us know so that we may include them in the Equipment/Materials list in the Instructor's Guide.

Write: Curriculum Revision and Development Department
National Center for Construction Education and Research
P.O. Box 141104, Gainesville, FL 32614-1104

Fax: 352-334-0932

E-mail: curriculum@nccer.org

Craft ______ Module Name ______

Copyright Date ______ Module Number ______ Page Number(s) ______

Description ______

______

______

______

(Optional) Correction ______

______

______

(Optional) Your Name and Address ______

______

______

Module 62202-02

Facility Inspection
(CT 27)

Introduction to the Pipeline Operator Qualification Rule, API, and NCCER

It could be said that petroleum products keep America running. Petroleum products like oil and natural gas are the principal sources of energy in the United States and are used in many manufacturing and chemical processes. Petroleum products are primarily transported by pipeline. There are more than a million miles of liquid and natural gas pipeline in this country, ranging in diameter from two inches to four feet. These pipelines bring crude oil and natural gas from production fields to processing plants and refineries—and ultimately to the consumer.

Pipeline Operator Qualification

Petroleum products are considered hazardous, so pipeline personnel must take extra precautions and follow specific procedures when working with them. To help ensure the safe handling and transportation of petroleum products, the United States Department of Transportation (DOT) mandates safe practices as federal regulations.

One such regulation, the Operator Qualification (OQ) rule, requires pipeline operators to develop a written qualification plan to evaluate their personnel and their contract personnel (see *49 CFR Part 192 Subpart N—Gas*, and *49 CFR Part 195 Subpart G—Liquid*). The evaluation hinges on the operator's ability to perform certain covered tasks and to be able to recognize and respond to abnormal operating conditions that may be encountered while performing these covered tasks.

Covered Tasks

A pipeline performance task is deemed *covered* if the activity passes the following four-part test:

1. Is it performed on a pipeline facility?
2. Is it an operations or maintenance task?
3. Is it performed as a requirement of *49 CFR Part 195 (Liquid)* or *49 CFR Part 192 (Gas)*?
4. Does it affect the operation or integrity of the pipeline facility?

If the answer to all four questions is *yes*, then the activity is considered *covered* and must figure into the evaluation process of pipeline personnel.

American Petroleum Institute

The OQ rule and its related covered tasks provided the catalyst to develop a unified pipeline training and assessment program—so that pipeline operators and owners have ready access to the assessment and training tools they need. This effort was the direct result of the vision and leadership provided by The American Petroleum Institute (API) and API members.

Efforts to unify the industry are nothing new to API. Indeed, the impetus for forming the Institute in 1919 was the need to standardize engineering specifications for drilling and production equipment. Today, the most pressing issues revolve around public perceptions and government policies toward the industry. Speaking with one voice on these issues has become as essential as having interchangeable parts in the field. API's members determine the petroleum industry's positions on public policy as well as the standards that govern its day-to-day operations worldwide.

Consortium on Operator Qualification

API assembled a focus group whose primary purpose was to identify covered tasks and provide compliance recommendations to pipeline operators.

The Consortium on Operator Qualification (COOQ) became the guiding force for the identification and development of covered tasks for the liquid pipeline industry. Their work product, API publication 1161, *Guidance Document for the Qualification of Liquid Pipeline Personnel*, formed the basis for the resulting Pipeliner Training and Assessment Program.

The covered task list created in API publication 1161 has now evolved to include gas covered tasks and is maintained by NCCER. The most current covered task list can be downloaded at www.nccer.org.

National Center for Construction Education and Research

With the guidance and support of API, its members, and the Consortium on Operator Qualification, the National Center for Construction Education and Research (NCCER) was able to design and implement the Pipeliner Training and Assessment Program.

NCCER is a not-for-profit 501(c)(3) education foundation established in 1995 by the world's largest and most progressive construction and maintenance contractors and national contractor associations. NCCER was founded to address the

severe workforce shortage facing the construction and maintenance industries and to develop a standardized training process and curricula. Today, NCCER is supported by hundreds of leading construction and maintenance companies, manufacturers, and national associations and developed more than 35 construction and maintenance titles.

NCCER also maintains a National Registry, which contains the records of trainees and instructors who have received NCCER training. All completion information for every trainee is sent to the NCCER and filed within the Registry. The Registry can then confirm training and skills for workers as they move from company to company, state to state, or even within their own company.

Standardized Curriculum (NCCER's Contren™ Learning Series)

NCCER curricula are developed using the expertise of industry subject matter experts (SMEs). The use of SMEs as an integral part of every new development or revision ensures that the curriculum is current, industry-driven, and indicative of industry practices. The development and/or revision process is facilitated from start to finish by NCCER project managers, who work closely with the SMEs and a select team of technical writer(s) on the development and presentation of the curriculum.

Pipeline: Five Training Paths

The pipeline curriculum focuses on essential skills and competencies within five skill areas:

- Corrosion Control
- Electrical and Instrumentation
- Maintenance
- Mechanical
- Operations (Field and Control Center)

SMEs from the pipeline industry have determined the skills and competencies needed based on API's covered tasks and the consensus of the group/industry. Specific company information is generally not incorporated into the curriculum unless the SMEs determine that the information is the industry standard.

Training Modules and Covered Tasks

The smallest unit of instruction in the curriculum is a *module*. Each curriculum is composed of modules (the equivalent of stand-alone chapters) that are logically sequenced into complete *levels* of instruction. Every module is task-oriented and requires the successful completion of a written (knowledge-based) examination and a performance (hands-on) test. Modules may be individually selected and/or combined for efficient task training. Modules in the pipeline curricula, to the extent possible, will align with API-developed covered tasks. While not all modules will relate to covered tasks, all covered tasks will relate to at least one module.

Benefits of Training to the Pipeline Marketplace

Applying NCCER's standardized training process allows pipeline owners and contractors in the United States to establish the benchmark for acceptable skill and competency levels within the pipeline industry. This process will assist training organizations in developing and/or improving standardized pipeline training programs that effectively match their resources and operating conditions while responding to regulatory demands and to the needs of the industry. The process will also recognize training organizations that meet the minimum standards of competence, warranting regulatory, public, and professional confidence. This in turn will foster national unity within the pipeline industry for the mutual benefit of individual technicians, owners, contractors, and ultimately pipeline-users and society.

Additional Resources

To review the Pipeliner Training and Assessment Program Implementation and Resource Guide and the latest version of pipeline covered tasks, visit NCCER's Web site at www.nccer.org, or contact:

National Center for Construction Education and Research
PO Box 141104
Gainesville, FL 32614-1104
(888) 622-3720
Web site: www.nccer.org

For more information on API, please contact:

American Petroleum Institute
1220 L Street, Northwest
Washington, D.C. 20005
(202) 682-8000
Web site: www.api.org

To order *API Pub 1161*, contact Global Engineering Documents by calling 1-800-854-7149 or visiting their Web site at www.global.his.com.

Additional sources of information on the pipeline operator qualification regulation include:

DOT – Office of Pipeline Safety
www.ops.dot.gov

Transportation Safety Institute
www.tsi.dot.gov

COURSE MAP

This course map shows all of the modules in the second level of the Pipeline Maintenance curriculum. The suggested training order begins at the bottom and proceeds up. Skill levels increase as you advance on the course map. The local Training Program Sponsor may adjust the training order.

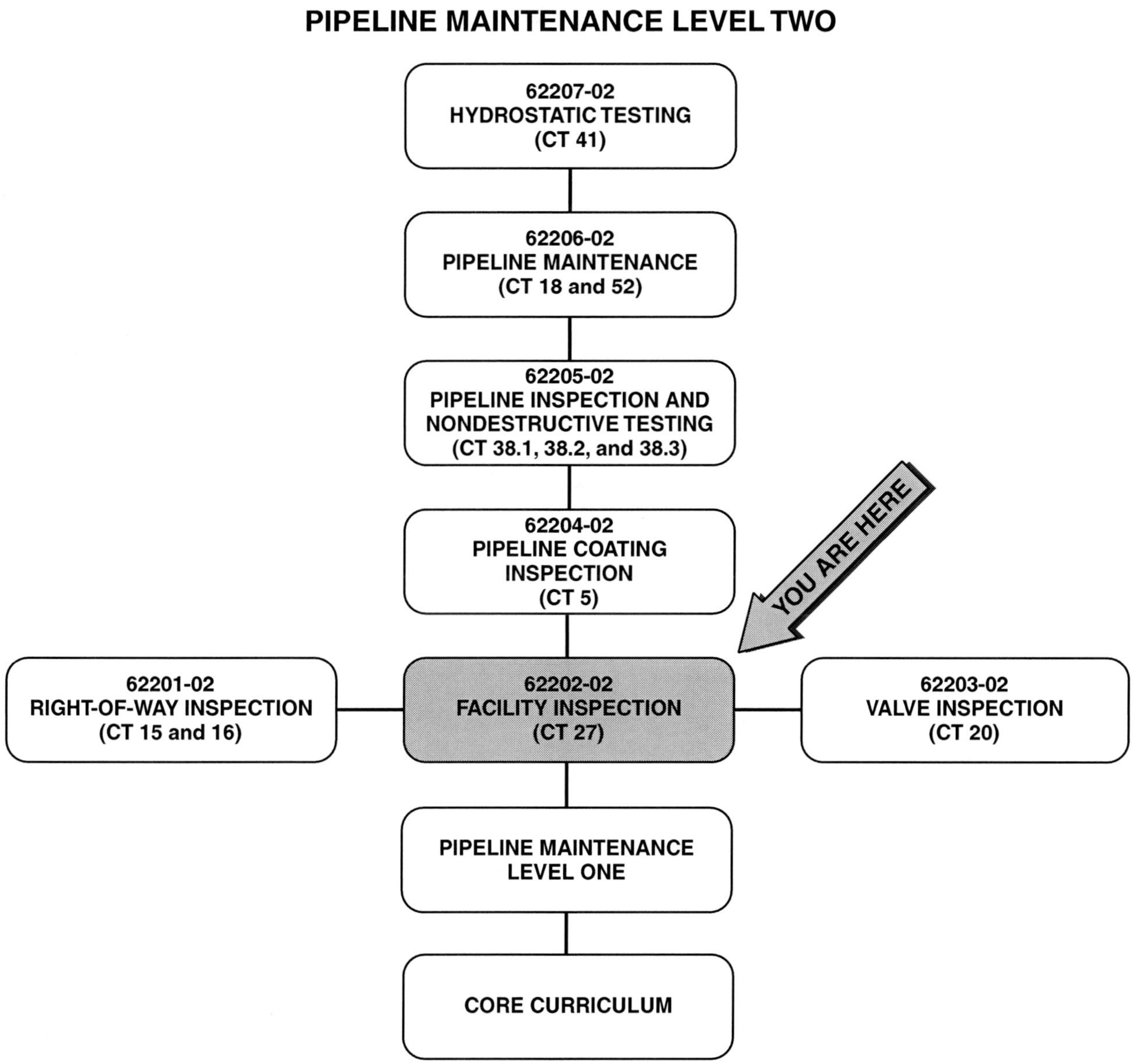

202CMAP.EPS

MODULE 62202-02 CONTENTS

Figure

MODULE 62202-02

Facility Inspection (CT 27)

Objectives

When you have completed this module, you will be able to do the following:

1. Inspect facilities (buildings).
2. Perform routine monthly inspections of breakout tanks (CT 27.1).
3. Perform inspections of in-service breakout tanks (CT 27.2).
4. Perform vault inspection and maintenance.

Prerequisites

Before you begin this module, it is recommended that you successfully complete the following: Core Curriculum; Pipeline Core, Modules 66101-02 and 66102-02; Pipeline Maintenance Level One.

Required Trainee Materials

1. Pencil and paper
2. Appropriate personal protective equipment

1.0.0 ◆ INTRODUCTION

Periodic inspections of all pipeline systems and facilities is necessary for maintenance and is required by Department of Transportation (DOT) regulations. Routine inspections are necessary to ensure that properties are kept up to standards of safety and to discourage vandalism and crime. Questionable items are noted, repairs are scheduled, and appropriate reports and documentation are submitted.

The inspection of facilities includes checking the operation of mechanical and electrical systems, physical conditions, and security effectiveness and recognizing hazardous or abnormal conditions.

This module covers:

- Visual inspections
- Security inspections
- Building inspections
- Tank farms and tank inspections
- Vault inspections and maintenance

2.0.0 ◆ VISUAL INSPECTIONS

Regularly scheduled visual inspections are necessary and required by companies to see that facilities are kept in proper condition. In the terminal, the inspection is performed by designated personnel. In the field, the station operator is usually designated to inspect and possibly perform repairs. Regular visual inspections of facilities include checking the general appearance, examining machinery, examining physical property conditions, and using personal senses.

2.1.0 General Appearance

The general appearance of a facility is checked from the outside upon arrival. This inspection includes checking the appearance of buildings and installations and looking for evidence of vandalism or trespassing. The condition of fences, gates, locks, signs, and other security measures is also checked along with the cleanliness of the company area.

2.2.0 Mechanical Examination

Inside the installation, the inspecting employee checks the condition of all pipes, pumps, valves, and other equipment and looks for any indications of a leak. Damaged equipment and any appearance of rust or corrosion is documented.

2.3.0 Physical Property Conditions

The visual inspection of physical property conditions includes checking the condition of buildings, windows, roofs, railings, stairs, floors, and other construction. Always ensure proper housekeeping within buildings and installation. Proper storage of parts, tools, and equipment is also checked.

2.4.0 Personal Senses

Along with the obvious appearance items, the employee uses personal senses to detect:

- Loud noises indicating improperly operating or failing equipment
- Excessive heat produced by a bearing, motor, or other equipment indicating failure
- Strong odors indicating product vapors leaking or equipment failure

3.0.0 ◆ SECURITY SYSTEM INSPECTIONS

Each pumping station, breakout tank, and other exposed facility must be provided with protection from vandalism and unauthorized entry. At a minimum, pump stations and breakout tanks are secured with standard fencing and locks. Security systems also include motion detectors, cameras, lighting, and alarms in addition to guards. Each facility's security measures are evaluated at least annually, and reliability of the system depends on monitoring, regular inspections, and repairs of the system.

Each pumping station, breakout tank area, terminal, and storage facility displays a sign visible to the public. Signs must be appropriately sized and laid out to prominently display the company name and the emergency telephone number. Signs must be periodically inspected to ensure that they are in good condition, secure, and legible.

Perform the following steps to inspect the pipeline facility security system:

Step 1 Deactivate the security alarm prior to entering the facility, or notify appropriate personnel prior to entering a remotely monitored facility.

Step 2 Inspect the facility perimeter fencing, gates, or other entry barriers for integrity and/or possible damage.

Step 3 Inspect all locks for possible damage, and ensure that locks are in working order.

Step 4 Inspect signs for proper placement, for visibility to the public, and to ensure that the information on the signs is correct and legible.

Step 5 Inspect local monitoring devices, such as motion detectors, alarms, and cameras, for operability, damage, or required maintenance.

Step 6 Test intrusion devices to ensure operability.

Step 7 Perform repairs to security equipment as appropriate, or prepare a facility repair order.

Step 8 Reactivate the facility security system, or notify appropriate personnel upon leaving the facility.

4.0.0 ◆ BUILDINGS

A variety of buildings may be included in pipeline facility inspections. These include the control center pump stations, booster stations, living quarters, and terminals. Each facility has specific mechanical and equipment inspection requirements, but the buildings themselves require attention to the following:

- Safe and orderly operating conditions
- Fire safety
- Earthquake safety
- Electrical safety
- Chemical storage
- Hazardous wastes
- Compressed gases

The following are checklists that can be used in performing these inspections.

4.1.0 Operating Conditions

When inspecting for proper operating conditions, you should check the building for the following:

- All areas are clean and orderly.
- There are no tripping hazards, wet spots, grease/oils, protruding objects, or miscellaneous debris.
- Permanent aisles/passageways where forklifts are used are properly marked.

- Open pits, tank ditches, etc., are covered or provided with standard guard rail protection.
- Load limits for stored items in elevated areas are clearly marked.
- Food and beverages are stored only in refrigerators specifically marked for these items.
- Areas where food and drink are prohibited are clearly marked.
- Unused equipment is stored in a safe and orderly manner.
- Warning and hazard signs are posted where required.
- Exterior building or trailer identification numbers are posted for emergency response.
- Emergency exits are correctly marked, visible, and accessible and have a minimum width of 28 inches.
- Doors that are not exits but could be mistaken for exits are clearly marked *Not an Exit*.
- There is sufficient lighting.
- There is sufficient ventilation.
- The noise level permits normal working conversation and safe communication.
- Unguarded holes or openings in floors are properly covered.
- Elevated platforms and working areas have standard rails.
- Elevated platforms and working areas have toe boards.
- Fixed industrial stairs are in good condition.
- Essential eye-wash stations are flushed weekly and documented.
- Emergency showers are working.

4.2.0 Fire Safety

When inspecting for fire safety, you should check the building for the following:

- Fire extinguishers are clearly accessible, have seals intact, and are properly mounted to the wall.
- The number of fire extinguishers required for the facility is provided, and the fire extinguishers are suitably located.
- The backs of inspection tags on fire extinguishers have been initialed and dated by persons responsible for monthly checks.
- Fire sprinklers have a minimum of 18-inch clearance below, are clear of interference by stored materials, and are clear of paint on heads.
- The presence of combustible materials is minimized.
- Flammable and combustible liquids are properly labeled and stored.
- The emergency exits are correctly marked, visible, and accessible and have a minimum width of 28 inches.
- All exit/access corridors are free of storage.

4.3.0 Earthquake Safety

When inspecting for earthquake safety, you should check the building for the following:

- All heavy objects are not on shelves higher than 5 feet, and all shelves are below 7 feet 6 inches.
- Furniture and equipment that could tip and block an exit are properly anchored.
- Stored materials are stacked securely to prevent tipping and scattering, which may cause tripping.
- Heavy workbenches are anchored.
- Shelf units are bolted to a wall, floor, or desk and are not ceiling high.
- Books on shelves are restrained.
- Computers and copying equipment are held firmly in place.
- Wheels on large machines are properly blocked.
- Machine shop fixed equipment and electronic racks are anchored.
- Air conditioners are properly secured and cable trays are braced at intervals of 10 feet or less.

4.4.0 Electrical Safety

When inspecting for electrical safety, you should check the building for the following:

- Extension cords:
  - Are not used in place of permanent wiring
  - Are not run through walls, doors, or ceilings
  - Are equipped with proper plugs
  - Use three-conductor cable
  - Have no damaged or taped ends
  - Are not daisy chained
- Power cords are not fraying, ground pins are in place, and necessary strain-relief measures have been taken.
- Frayed wires have been replaced.
- Access to electrical panels is clear and unobstructed for at least 36 inches.
- Access to switches and circuit breakers is clear and unobstructed.

- All receptacles located within 6 feet of sinks and exterior doors, on all metal-surfaced workbenches, and outside are ground-fault circuit interrupters (GFCIs).
- Protective covers are in place over boxes, raceways, and fittings.
- Power cabinets and breakers are properly labeled.
- All energized parts are guarded to prevent accidental contact.
- Electrical panel directories are in place and accurate.
- The door or panel that encloses the panel box is either bolted or locked closed.
- The access door to the panel breakers or fuses is unlocked to allow access.

4.5.0 Chemical Storage

When inspecting for the proper storage of chemicals, you should check the building for the following:

- Cylinder bottles or six packs are properly restrained.
- Container caps are in place when not in use.
- Tags are attached showing whether a container is full, empty, or in use.
- Cylinders are clearly marked as to their contents.
- Cabinets and containers are properly labeled.
- There are separate disposal areas or containers for rags, glass, and trash.
- Bottles, cans, vials, and flasks are properly labeled with contents and date.
- Flammables are stored in the appropriate cabinet with spill prevention and containment provided.
- Organics, acids, and bases are stored apart from each other.
- Spill-containment materials are readily available for simple or small spills.
- Appropriate containers are emptied daily.
- Material safety data sheets (MSDSs) are available for all hazardous substances present.

4.6.0 Hazardous Wastes

When inspecting for the proper containment of hazardous wastes, you should check the building for the following:

- A waste accumulation area (WAA) is designated.
- Appropriate disposal cans are available, properly labeled with contents and date, and clearly marked *Hazardous Waste*.
- The accumulation start date and contact name are provided on containers.
- Containers are closed except when being filled or emptied.
- There is separate secondary containment for incompatible materials.
- There are separate waste containers for solvent solids waste and oil and coolant solids waste.
- There is a WAA logbook on hand.

4.7.0 Compressed Gases

When inspecting for the proper containment of compressed gases, you should check the building for the following:

- Cylinder bottles or six packs are properly restrained.
- Container caps are in place when not in use.
- Tags are attached showing whether the container is full, empty, or in use.
- Cylinders are clearly marked as to their contents.

5.0.0 ◆ TANK FARMS AND TANKS

Tank farms and the tanks within them hold enormous amounts of petroleum products. These stores could produce great damage if an incident occurred. Therefore, it is critical that tank farms and tanks be inspected carefully and often.

5.1.0 Tank Farms

Tank farms must be inspected and documented according to company policy. The following guidelines detail some areas of concern:

- Observe all other system lines, valves, and controls for leaks and proper operation.
- Inspect electrical equipment for damaged or loose conduit, junction box covers that are out of place, exposed wiring, and any lights that are not operating.
- Verify that dike drain valves are closed and locked.
- Drain water from dikes as needed, and document on the appropriate form.
- Note any fencing repair and trash cleanup needed.
- Note any sign of vandalism, forced entry, or theft.
- Check for excessive growth of vegetation.
- Check the secondary liners or containers in the tank.
- Report any abnormal condition to the shift supervisor, and prepare any documentation required.

5.2.0 Tanks

Tanks are inspected externally on monthly and annual schedules for the following:

- Stability of the foundation
- Condition of the shell, stairs, and roof
- Condition of the venting or safety valve equipment
- Condition of the firewalls or dikes
- Condition of the seals, including internal deck seals
- Condition of the gauging equipment
- Condition of the mixers and electrical equipment
- Condition of grounding
- Condition of the **roof drains**
- Condition of the level alarms

A visual inspection is performed, and any problems are recorded and corrected as necessary. The tank is inspected internally when practical.

Inspection results are recorded on the tank inspection report form with the internal inspection noted if performed. The form asks for the location of the tank and a description of the condition of each of the physical parts of the tank and all of the associated equipment. Repairs that are required and/or made must also be described. The form is dated and signed by the inspector. *Figure 1* shows a tank inspection report form.

5.2.1 Monthly Inspections

Monthly inspections of a tank are performed on a routine, scheduled basis that is consistent with conditions at the particular site; the scheduled interval is not to exceed one month. Monthly inspections are performed in accordance with *API Standard 653 – Tank Inspection, Repair, Alteration, and Reconstruction*. The purpose of the task is to visually inspect the tank's exterior surface and tank components to identify any potential problems, faulty equipment, or conditions that could result in a potential safety or environmental hazard. The steps for performing a monthly inspection are:

Step 1 Visual inspection of the foundation is performed to identify abnormal conditions, such as settlement, underside corrosion, or leads and to determine foundation levelness and bottom elevations. Visually inspect the foundation for defects such as:

- Broken concrete, spalling, and cracks
- Indications of bottom leakage from the drain openings in the foundation ring, the back of water draw basins, and the top surface of the ring
- Cavities under the foundation
- Vegetation against the bottom of the tank
- Problems with rainwater draining from the shell away from the tank
- Settlement around the perimeter of the tank

CAUTION

Any settlement noted in the foundation of a tank may indicate active ground movement or insufficient sub-grade strength. Continued settlement could result in damage to the tank shell or could result in tank failure. Document the settlement location, and estimate the magnitude of the settling or displacement and the impacts of such, if any, on the tank shell or equipment. Photographs are recommended to document further settlement with time. Report the condition to the appropriate company personnel for further inspection or repair.

Step 2 Visual inspection of the shell is performed to identify leaks, paint failure, pitting, and corrosion. Visually inspect the shell for defects such as:

- Paint failures, pitting, and corrosion
- Rivet and seam leaks
- Corrosion damage on the **wind girder** and handrail, especially at tack-welded junctions
- Broken welds

Step 3 Document any leak locations by sketch or photograph because the physical location will be lost when the tank is cleaned for repainting.

Step 4 Visual inspection of shell appurtenances is essential in identifying leaks, component failure, and corrosion. Inspect the shell components for defects such as:

- Cracks or signs of leakage on weld joints at nozzles, manways, and reinforcing plates
- Flange leaks and leaks around bolting
- Leaks from manifold piping, flanges, and valves
- Absence of or problems with fire-fighting system components
- Leaks from sample connections
- Valve operation problems
- Leaks from the autogauge tape guide and lower sheave housing
- A damaged autogauge head
- Problems with the autogauge head checker tape movement
- Lack of freedom of movement of the marker and float

MONTHLY TANK INSPECTION REPORT

Date: Company: Station:

Tank #: ______________ Capacity: ______________ Inspector: ______________

Location: ______________________________

Tank Dimensions ______________________________

TANK INFORMATION		Yes	No	N/A	Remarks
Tank Construction:	Welded				
	Bolted				
Tank Type:	Fixed / Coned				
	IFR				
	EFR				
Foundation Type:	Asphalt				
	Sand Base				
	Concrete / Ringwall				
	Tank Band				
	Crushed Rock				
Tank Foundation / Bottom Observations:	Signs of Bottom Leakage				
	Is the shell to bottom joint visible				
	Signs of Settlement around the Perimeter Tank				
	Is the foundation damaged				
Tank Shell Observations:	Shell Coating failure / pitting / corrosion				
	Shell to bottom weld area have corrosion defects or thinning				
	Signs of shell leakage / corrosion				
	Tank shell insulated				
	Condition of Insulation				
	Shell Distortions				
Tank Piping / Appurtenances:	Stress,strain, or shifted piping				
	Signs of corrosion, defects around manways / nozzles / attachments				
	Signs of flange leaks and leaks around bolts				
	Signs of leaks, thermal relief , damage on piping and valves				
	Are gauging devices operational				
	Is there adequate reinforcement around nozzles and appurtenances				
	Does the mixer have adequate supports				
	Does the mixer shows signs of leakage or defects				
Fixed Roof Inspection	Is the fixed roof insulated				
	Is the roof insulation in good condition				
	Are there low spots on the fixed roof				
	Does water accumulate on the roof / insulation				
	Are roof nozzles / manways in good condition				
	Roof coating failure				

Comments:

202F01.EPS

Figure 1 ◆ Tank inspection report form.

> **WARNING!**
> A component or equipment failure could pose a safety or environmental risk to employees, the public, or the property adjacent to a breakout tank. If it is safe to do so, close, isolate, or remove the failed component from operation. Immediately notify the operations control center and appropriate company personnel. Document the condition for further inspection or repair. If the failed component presents an imminent hazard condition, evacuate the area immediately, and notify the emergency response team.

Step 5 Visual inspection of the tank roof will assist in identifying problems with paint failure, corrosion, standing water, and a non-level roof. It also helps to identify any excessive gaps between the shell and the seals on the tanks that may be subject to air quality control regulations. Inspect the tank roof for defects such as:

- Paint failure, holes, pitting, and corrosion on the roof deck
- Indications of standing water caused by inadequate drainage, an unlevel roof, or leaking **pontoons**
- Sagging of the fixed-roof deck indicating potential rafter failure
- Deterioration, holes, tears, and cracks in the fabric of the roof seal system
- Openings in the seals that would permit vapor emissions
- Protruding bolt or rivet heads against the shell
- Incorrect positioning of the pontoon covers on the external roof

Step 6 Tank roof appurtenances should be inspected for proper operation and maintenance. Inspect the roof appurtenances for defects such as:

- Corrosion or other problems on the sample hatch cover
- Corrosion or missing bolts on the autogauge inspection hatch
- Corrosion on the autogauge float well cover
- Obstructions of the wheels and run ners that could cause the ladder to get off-track and damage the roof
- Significant damage to and cracked or broken welds in the stair treads, handrails, and supports

Step 7 Inspect the area for buildup of trash, vegetation, and other flammable materials.

Step 8 Remove any debris, vegetation, or flammable materials that could contribute to a fire or create a safety hazard.

Step 9 Document the monthly tank inspection on the appropriate company report form. The report should identify items that were visually inspected and items that need immediate attention.

5.2.2 5-Year In-Service Inspections

In-service inspections are performed in accordance with *API Standard 653 – Tank Inspection, Repair, Alteration, and Reconstruction*. An authorized inspector certified under this standard must perform this task. The steps here are for in-service tank inspections; additional steps, also detailed under *Standard 653*, are required for out-of-service tanks. The steps for performing in-service inspections are:

Step 1 Inspect the foundation for defects such as:

- Broken concrete, spalling, and cracks, particularly under backup bars used in welding butt-welded annular rings under the shell
- Indications of bottom leakage from the drain openings in the ring, the backs of water draw basins, and the top surface of the ring
- Cavities under the foundation
- Vegetation against the bottom of the tank
- Settling of the tank into asphalt or other base that could direct runoff under the tank
- Areas where leaching of oil has left rock filler exposed, indicating hydrocarbon leakage
- Presence of crushed rock under the tank bottom
- Problems with draining away from the tank and associated piping and manifolds
- Problems with dike drain operations
- Buildup of debris, vegetation, and flammable materials

Step 2 Remove any debris, vegetation, and flammable materials.

Step 3 Inspect the shell for defects such as:

- Paint failures, pitting, and corrosion
- Corrosion on and thinning of the plate and weld of the bottom angle area
- Problems with the bottom-to-foundation seal

- Internal grooving, pitting, corrosion and coating failures on the floating roof tank
- Rivet and seam leaks
- Corrosion loss and wear of rivets
- Joint problems with vertical seams
- Paint failure, pitting, or corrosion build-up on the wind girder and handrail, especially at tack-welded junctions
- Broken welds
- Pitting on the support welds to the shell, especially shell plates

CAUTION
Pipeline system damage may be caused by corrosion. During an inspection, corrosion on the tank or tank components may be found. Document any corrosion location, and identify the affected portion of the tank roof, shell, or components. Photographs may help in assessing the corrosion. Report the condition to the appropriate company personnel for further inspection or repair.

Step 4 Sketch or photograph any leaks because the physical location will be lost when the shell is cleaned for painting.

Step 5 If no record of the vertical riveted seams exists, dimension and sketch or photograph the rivet pattern, including the number of rows, rivet size, pitch length, and whether it is butt or lap riveted.

Step 6 Note if the supports on the floating roof tank have reinforcing pads welded to the shell.

Step 7 Inspect the shell manways and nozzles for defects such as:

WARNING!
A fire or explosion may occur because of a release of vapors or liquid from the tank. Evacuate the area immediately, and notify local 911 emergency personnel and the company's emergency response team.

- Cracks or signs of leakage on weld joints at nozzles, manways, and reinforcing plates
- Shell plate dimpling around nozzles caused by excessive pipe deflection
- Flange leaks and leaks around bolting
- Problems with insulation sealing around the manways and nozzles
- Inadequate manway flange and cover thickness on the mixer manways

WARNING!
While inspecting a tank, an unexpected hazardous liquid leak may be discovered. Notify local emergency personnel if appropriate, make the proper company notifications, and try to identify the source of the leak if possible.

Step 8 Inspect the tank piping manifolds for defects such as:

- Leaks from the manifold piping, flanges, and valves
- Problems with fire-fighting system components
- Anchored piping that would be hazardous to the shell or bottom connections during earth movement
- Inadequate thermal pressure relief of the piping to the tank
- Inoperative regulators for tanks with purge gas systems
- Leaks from and problematic valve operation on sample connections
- Damaged or inaccurate temperature indicators
- Bad welds on shell-mounted davit clips above valves 6 inches and larger

Step 9 Inspect the autogauge system for defects such as:

- Leaks in the autogauge tape guide and lower sheave housing
- A damaged autogauge head
- Improper movement of the autogauge head checker tape
- A history of tape tending to hang up during floating roof movement
- An inaccurate level reading on the autogauge, with the maximum variation being 2 inches
- More than 2 feet of tape exposed at the end of the tape guide when the roof is in the lowest position
- Bad condition or illegibility of board-on-board autogauges
- Lack of freedom of movement of the marker and float

Step 10 Inspect the shell-mounted sample station for defects such as:

- Malfunctioning of valves and plugging of lines, including drain or return-to-tank lines
- Leaks and operating problems in the circulation pump
- Problems with the test bracing and supports of sample lines and equipment

Step 11 Inspect the shell manway mounted heater for the presence of oil in the condensate drain, which indicates leakage.

Step 12 Inspect the mixer for defects such as:

- Improper mounting flange and support
- Leakage
- Problems with the condition of the power lines and connections to the mixer

Step 13 Inspect the **swing lines** for defects such as:

- Improper lowering of the swing line, resulting in loose cable
- Improper movement of the indicator
- Leaks at threaded and flanged joints on the external guide system
- Significant difference in the specific gravity of the product, which identifies ballast varying need
- Corrosion over the entire length of the nonstainless steel cable
- Wear or fraying of the cable
- Water or gravity differences in product sample comparison, which indicates a leaking swing joint
- Problems with direction and height indicated on target

Step 14 Inspect the roof for defects such as:

- Deck plate internal corrosion
- Deck plate external corrosion
- Problems with deck drainage indicated by standing water
- Sagging of the fixed roof deck, which indicates potential rafter failure
- A nonlevel roof
- The presence of explosive gas on top of the internal floating roof
- Cracks or leaks in the roof insulation weather coat
- Wet insulation under the weather coat
- Corrosion or holes in the roof deck under the edge of the insulated area

Step 15 Inspect the floating roof seal systems for defects such as:

- Problems with gaps and annular space
- Seal fabric on the primary shoe seals pulling away from the shell
- Deterioration, holes, tears, or cracks in the fabric
- Corrosion and wear of visible metallic parts
- Openings in seals that would permit vapor emissions
- Protruding bolt or rivet heads rubbing against the sides of the shell
- Signs of buckling or shallow angle of the secondary seals
- Inflexibility, cracks, or tears in the wedge-type wiper seals

WARNING!

During the inspection of a breakout tank, you may encounter seals that are rolled under. Improperly installed or damaged tank roof seals may allow a liquid or vapor release from the tank. Document the seal condition, location, and affected portion of the tank roof. Report the condition to the appropriate company personnel for further inspection or repair.

Step 16 Inspect the sample hatch for defects such as:

- Malfunctioning or bad condition of the cover
- Bad condition of the seal inside the cover
- Corrosion on and plugging of the thief and gauge hatch cover
- Inaccurate marker and tab stating the hold-off distance, where used to reel gauge stock level
- A missing reinforcing pad where the sample hatch penetrates the roof deck
- Bad condition of or operation problems with the recoil reel and rope
- Faulty operation of the entire system
- Missing or bad condition of the protective coating or liner inside the sample hatch for ultra clean stocks

Step 17 Inspect the gauge well for defects such as:

- Thinning or bad condition of the gauge well
- Missing or illegible hold-off distance marker and tab
- Bad condition of the roof guide and grooved rollers on the floating roofs
- Problems with the gauge tank well pipe distance to the tank shell
- Leakage in the gauge well washer valve, if present, and a missing bull plug or blind flange

Step 18 Inspect other roof appurtenances for defects such as:

- Corrosion and wear on and structural problems with the scaffold support
- Corrosion or missing bolts on the autogauge inspection hatch and guides

- Corrosion on the tape guide's and float guide's wire anchors
- Corrosion on the float well cover
- Wear or fraying of the tape cable
- Problems with the condition and locking of the pivot pins on the roof-mounted vents
- Plugged drain return to the tank on the gauging platform drip rings
- Problems with the emergency roof drains
- Bad condition of the screen on the **rim vent** cover
- Plating or removal of the rim vents
- Bad settings on the **vacuum breakers**
- Leakage from the pontoons
- Explosive gas

> **WARNING!**
> Vapors or liquid may be released from the bottom plate, shell, or roof of the tank. A release may be encountered during an inspection. Notify local 911 emergency personnel if appropriate, make the proper company notifications, and try to identify the source of the leak if possible.

- Plugged-up vent tubes of pontoon hatches that are equipped with locked-down covers
- Improperly operating lock down devices on pontoon hatches

Step 19 Inspect the rolling ladder for defects such as:

- Corrosion on the stringers
- Bad weld attachment of the rungs to the stringers
- Corrosion where the rungs are welded to the stringers
- Wear and corrosion where the rolling ladder attaches to the gauging platform
- Wear and insecurity of the pivot bar
- Bad operation of the self-leveling stairway treads
- Corrosion and wear on the moving parts
- Restricted movement, flat spots, or axle wear on the ladder wheels
- Misalignment of the ladder with the roof rack
- Excessive wear of the ladder track caused by the wheels
- Corrosion of the track welds
- The track support reinforcing pads not seal-welded to the deck plate
- A missing hand rail on top of the shell if the ladder track extends to within 5 feet of the roof edge
- Incorrect angle of rolling ladder when the roof is on low legs

> **CAUTION**
> Faulty equipment/appurtenances, such as ladders, gauge hatches, mixers, and pontoons, are a safety hazard to company and contract personnel. Faulty or damaged equipment may cause a release from the tank. Equipment or tank appurtenances that are faulty or damaged should be documented on the tank inspection report. If the equipment presents an immediate safety hazard, the tank or tank location should be isolated, tagged, or cordoned off to prevent access to the area. Report the condition to the appropriate company personnel for further inspection or repair.

Step 20 Inspect the access structures for defects such as:

- Pitting, holes, or paint failure on the hand rails
- Bad attachment welds on the hand rails
- Cold joints and sharp edges on the handrails
- Corrosion on, malfunctioning of, or wrong length of the safety drop bar or chain
- A hazardous opening between the rolling ladder and the gauging platform when the floating roof is at its lowest level
- Corrosion and paint failure on the platform frame
- Corrosion on and weld failure of the attachment of the frame to the supports or supports to tank
- A lack of reinforcing pads where the supports are attached to the shell or roof
- Thinning of or holes on the surface where the deck plate or grating rests
- Bad or missing seal welds at flat surface to flat surface junctures
- Thinning of or holes and paint failure on deck plate
- Rust scale buildup on the plate-to-frame weld
- Corrosion-caused thinning and failure of welds on the grating
- Problems with the grating tie-down clips or stair risers
- Corrosion on, paint failure of, and weld failure of the spiral stairway stringers

- An insecure attachment of the stairway treads to the stringer
- Problems with the stairway supports connected to the shell welds and reinforcing pads
- Corrosion on the steel support attachment to the concrete base

6.0.0 ◆ TERMS AND DEFINITIONS

The following list highlights terms that are commonly used during pipeline tank inspections. It is important to be familiar with these terms and their definitions:

- *Authorized inspector* – A person certified as an aboveground storage tank inspector per *API Standard 653, Appendix D.*
- *Deck legs* – Steel pipes mounted on the floating roof that provide support for the roof when the liquid is drained below a set level. Deck legs are typically adjustable to allow for a change in supported roof height from the bottom of the tank. They are often referred to as high or low legs.
- *Geodesic domes* – Self-supporting dome roofs (typically aluminum) on tanks that are intended to prevent rain water from entering the tanks.
- *Guide pole* – Vertical pole device that extends through the floating roof of a tank to maintain the roof in a centered position and to prevent the roof from rotating.
- *Overflow drain* – Nozzle on the shell of a tank located approximately one foot below the top of the tank. Overflow drains are always larger than tank inlet nozzles.
- *Primary and secondary seals* – Flexible sealing devices mounted on the floating roofs of tanks to provide a reasonably close fit between the roofs and the internal walls of the tanks. The primary seal is typically made of fabric and is mounted to be in contact with the liquid in the tank. The secondary seal is mounted on top of the tank roof.
- *Sketch plate* – The bottom plate on which a tank shell rests.

Summary

Performing periodic inspections of pipeline systems and facilities is very important for maintenance. This module discusses the different types of visual inspections. These inspections include checking for general appearance, performing a mechanical inspection, checking the physical property conditions, and using your senses. The different types of installation inspections are also covered. The inspections described in this module are for security systems, buildings, tank farms, and tanks. The differences between monthly and annual inspections for tanks are also discussed in this module.

Review Questions

1. Periodic inspections of all pipeline systems and facilities is required by _____ regulations.
   a. ANSI
   b. DOT
   c. state
   d. regional

2. Regular visual inspections of facilities include all of the following *except* _____.
   a. checking the general appearance
   b. examining machinery
   c. using personal sense
   d. testing equipment

3. Each facility's security measures are evaluated at least _____.
   a. once a month
   b. twice a year
   c. once a year
   d. once every two years

4. Each pumping station, breakout tank area, terminal, and storage facility displays a sign visible to the public that prominently displays the company name and _____.
   a. emergency telephone number
   b. type of product being pumped/stored
   c. warnings
   d. company contact information

5. Tanks are inspected monthly and _____.
   a. weekly
   b. bimonthly
   c. biannually
   d. annually

6. Monthly inspections are made in accordance with _____ regulations.
   a. ANSI
   b. state
   c. API
   d. DOT

7. Any settlement noted in the foundation of a tank may indicate _____.
   a. porous soil
   b. active ground movement
   c. poor tank construction
   d. tank overload

8. Per *API Standard* _____, an authorized inspector is required to perform tank inspections.
   a. *550*
   b. *623*
   c. *653*
   d. *675*

9. Improperly installed or damaged roof tank _____ may allow a liquid or vapor release.
   a. decks
   b. seals
   c. float well covers
   d. pontoons

10. A _____ extends through the floating roof of a tank to maintain the roof in a centered position.
    a. guide pole
    b. deck leg
    c. sketch plate
    d. swing line

GLOSSARY

Trade Terms Introduced in This Module

Pontoons: External or internal compartments on a floating roof that provide buoyancy, allowing the roof to float.

Rim vents: Mechanical vents mounted on the floating roof of a tank that vent the vapor space between the primary and secondary rim seals.

Roof drains: Steel pipe or hose mounted on a floating roof to drain rain water.

Swing lines: Cables attached to the internal piping of the outlet connection of a tank.

Vacuum breakers: Valves mounted on top of the tank roof to prevent a vacuum under the roof.

Wind girders: Stiffening rings mounted on open top tanks to maintain shell roundness against wind loads.

CONTREN™ LEARNING SERIES — USER UPDATES

The NCCER makes every effort to keep these textbooks up-to-date and free of technical errors. We appreciate your help in this process. If you have an idea for improving this textbook, or if you find an error, a typographical mistake, or an inaccuracy in NCCER's Contren™ textbooks, please write us, using this form or a photocopy. Be sure to include the exact module number, page number, a detailed description, and the correction, if applicable. Your input will be brought to the attention of the Technical Review Committee. Thank you for your assistance.

Instructors – If you found that additional materials were necessary in order to teach this module effectively, please let us know so that we may include them in the Equipment/Materials list in the Instructor's Guide.

Write: Curriculum Revision and Development Department
National Center for Construction Education and Research
P.O. Box 141104, Gainesville, FL 32614-1104

Fax: 352-334-0932

E-mail: curriculum@nccer.org

Craft ______________________ Module Name ______________________

Copyright Date ____________ Module Number ____________ Page Number(s) ____________

Description

____

____

____

____

(Optional) Correction

____

____

____

(Optional) Your Name and Address

____

____

____

Module 62203-02

Valve Inspection
(CT 20)

Introduction to the Pipeline Operator Qualification Rule, API, and NCCER

It could be said that petroleum products keep America running. Petroleum products like oil and natural gas are the principal sources of energy in the United States and are used in many manufacturing and chemical processes. Petroleum products are primarily transported by pipeline. There are more than a million miles of liquid and natural gas pipeline in this country, ranging in diameter from two inches to four feet. These pipelines bring crude oil and natural gas from production fields to processing plants and refineries—and ultimately to the consumer.

Pipeline Operator Qualification

Petroleum products are considered hazardous, so pipeline personnel must take extra precautions and follow specific procedures when working with them. To help ensure the safe handling and transportation of petroleum products, the United States Department of Transportation (DOT) mandates safe practices as federal regulations.

One such regulation, the Operator Qualification (OQ) rule, requires pipeline operators to develop a written qualification plan to evaluate their personnel and their contract personnel (see *49 CFR Part 192 Subpart N—Gas,* and *49 CFR Part 195 Subpart G—Liquid*). The evaluation hinges on the operator's ability to perform certain covered tasks and to be able to recognize and respond to abnormal operating conditions that may be encountered while performing these covered tasks.

Covered Tasks

A pipeline performance task is deemed *covered* if the activity passes the following four-part test:

1. Is it performed on a pipeline facility?
2. Is it an operations or maintenance task?
3. Is it performed as a requirement of *49 CFR Part 195 (Liquid)* or *49 CFR Part 192 (Gas)*?
4. Does it affect the operation or integrity of the pipeline facility?

If the answer to all four questions is *yes,* then the activity is considered *covered* and must figure into the evaluation process of pipeline personnel.

American Petroleum Institute

The OQ rule and its related covered tasks provided the catalyst to develop a unified pipeline training and assessment program—so that pipeline operators and owners have ready access to the assessment and training tools they need. This effort was the direct result of the vision and leadership provided by The American Petroleum Institute (API) and API members.

Efforts to unify the industry are nothing new to API. Indeed, the impetus for forming the Institute in 1919 was the need to standardize engineering specifications for drilling and production equipment. Today, the most pressing issues revolve around public perceptions and government policies toward the industry. Speaking with one voice on these issues has become as essential as having interchangeable parts in the field. API's members determine the petroleum industry's positions on public policy as well as the standards that govern its day-to-day operations worldwide.

Consortium on Operator Qualification

API assembled a focus group whose primary purpose was to identify covered tasks and provide compliance recommendations to pipeline operators.

The Consortium on Operator Qualification (COOQ) became the guiding force for the identification and development of covered tasks for the liquid pipeline industry. Their work product, API publication 1161, *Guidance Document for the Qualification of Liquid Pipeline Personnel,* formed the basis for the resulting Pipeliner Training and Assessment Program.

The covered task list created in API publication 1161 has now evolved to include gas covered tasks and is maintained by NCCER. The most current covered task list can be downloaded at www.nccer.org.

National Center for Construction Education and Research

With the guidance and support of API, its members, and the Consortium on Operator Qualification, the National Center for Construction Education and Research (NCCER) was able to design and implement the Pipeliner Training and Assessment Program.

NCCER is a not-for-profit 501(c)(3) education foundation established in 1995 by the world's largest and most progressive construction and maintenance contractors and national contractor associations. NCCER was founded to address the

severe workforce shortage facing the construction and maintenance industries and to develop a standardized training process and curricula. Today, NCCER is supported by hundreds of leading construction and maintenance companies, manufacturers, and national associations and developed more than 35 construction and maintenance titles.

NCCER also maintains a National Registry, which contains the records of trainees and instructors who have received NCCER training. All completion information for every trainee is sent to the NCCER and filed within the Registry. The Registry can then confirm training and skills for workers as they move from company to company, state to state, or even within their own company.

Standardized Curriculum (NCCER's Contren™ Learning Series)

NCCER curricula are developed using the expertise of industry subject matter experts (SMEs). The use of SMEs as an integral part of every new development or revision ensures that the curriculum is current, industry-driven, and indicative of industry practices. The development and/or revision process is facilitated from start to finish by NCCER project managers, who work closely with the SMEs and a select team of technical writer(s) on the development and presentation of the curriculum.

Pipeline: Five Training Paths

The pipeline curriculum focuses on essential skills and competencies within five skill areas:

- Corrosion Control
- Electrical and Instrumentation
- Maintenance
- Mechanical
- Operations (Field and Control Center)

SMEs from the pipeline industry have determined the skills and competencies needed based on API's covered tasks and the consensus of the group/industry. Specific company information is generally not incorporated into the curriculum unless the SMEs determine that the information is the industry standard.

Training Modules and Covered Tasks

The smallest unit of instruction in the curriculum is a *module*. Each curriculum is composed of modules (the equivalent of stand-alone chapters) that are logically sequenced into complete *levels* of instruction. Every module is task-oriented and requires the successful completion of a written (knowledge-based) examination and a performance (hands-on) test. Modules may be individually selected and/or combined for efficient task training. Modules in the pipeline curricula, to the extent possible, will align with API-developed covered tasks. While not all modules will relate to covered tasks, all covered tasks will relate to at least one module.

Benefits of Training to the Pipeline Marketplace

Applying NCCER's standardized training process allows pipeline owners and contractors in the United States to establish the benchmark for acceptable skill and competency levels within the pipeline industry. This process will assist training organizations in developing and/or improving standardized pipeline training programs that effectively match their resources and operating conditions while responding to regulatory demands and to the needs of the industry. The process will also recognize training organizations that meet the minimum standards of competence, warranting regulatory, public, and professional confidence. This in turn will foster national unity within the pipeline industry for the mutual benefit of individual technicians, owners, contractors, and ultimately pipeline-users and society.

Additional Resources

To review the Pipeliner Training and Assessment Program Implementation and Resource Guide and the latest version of pipeline covered tasks, visit NCCER's Web site at www.nccer.org, or contact:

National Center for Construction Education and Research
PO Box 141104
Gainesville, FL 32614-1104
(888) 622-3720
Web site: www.nccer.org

For more information on API, please contact:

American Petroleum Institute
1220 L Street, Northwest
Washington, D.C. 20005
(202) 682-8000
Web site: www.api.org

To order *API Pub 1161*, contact Global Engineering Documents by calling 1-800-854-7149 or visiting their Web site at www.global.his.com.

Additional sources of information on the pipeline operator qualification regulation include:

DOT – Office of Pipeline Safety
www.ops.dot.gov

Transportation Safety Institute
www.tsi.dot.gov

COURSE MAP

This course map shows all of the modules in the second level of the Pipeline Maintenance curriculum. The suggested training order begins at the bottom and proceeds up. Skill levels increase as you advance on the course map. The local Training Program Sponsor may adjust the training order.

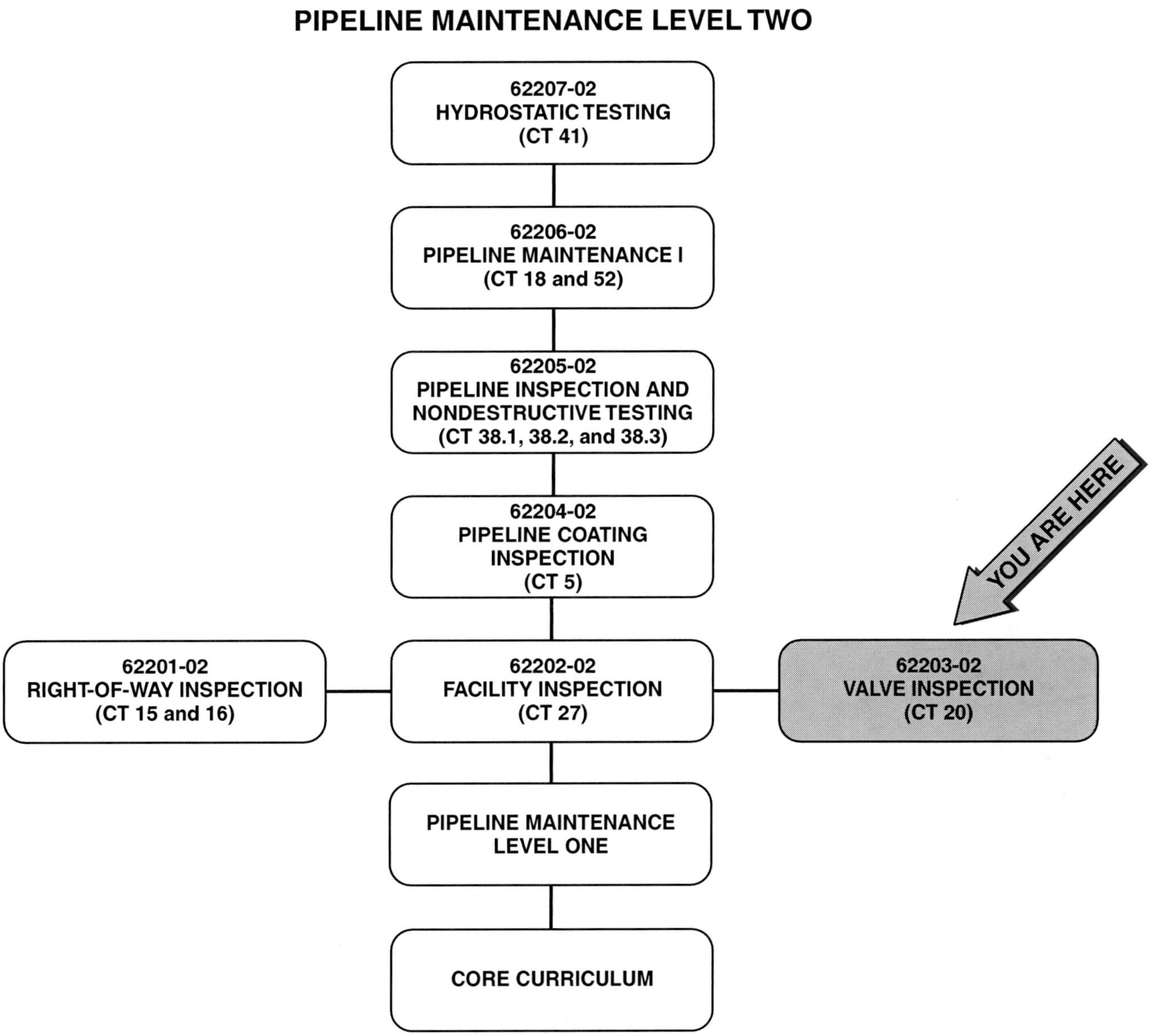

203CMAP.EPS

MODULE 62203-02 CONTENTS

Figures

Table

Valve Inspection (CT 20)

Objectives

When you have completed this module, you will be able to do the following:

1. Identify the types of valves used to start and stop flow on a pipeline.
2. Identify the types of valves used to regulate flow on a pipeline.
3. Identify and explain the functions of the various types of relief valves.
4. Perform a routine walk-around valve inspection and an external integrity inspection (CT 20.1 and 20.2).
5. Function test and leak test a valve (CT 20.3 and 20.4).
6. Perform preventative maintenance on pipeline valves.
7. Perform regular lubrication of valves, and winterize valves.
8. Install threaded and flanged valves.

Prerequisites

Before you begin this module, it is recommended that you successfully complete the following: Core Curriculum; Pipeline Core, Modules 66101-02 and 66102-02; Pipeline Maintenance Level One.

Required Trainee Materials

1. Pencil and paper
2. Appropriate personal protective equipment

1.0.0 ◆ INTRODUCTION

Valves are the basic control devices of the pipeline system. They are mechanical devices that start and stop product flow, control the rate of flow, and act as safety devices on the pipeline. Some are manually operated, while others are automatic. The entire system is controlled by the method, sequence, and timing of the opening and closing of respective valves.

Although there are many variations for particular applications, basic valve design is relatively uniform. All valves have two common characteristics. Every valve has a passageway in the **body** through which fluid passes and a moveable part that moves into the passageway to open, close, or change the size of the passageway. A valve can be used as a shutoff device for on-off service, a **throttling** device for controlling flow, a pressure-relief device for protecting the pipeline, or a safety device to prevent reversal of flow.

Valves are rated for maximum operating pressure (MOP) or maximum allowable operating pressure (MAOP). Valve ratings are provided by the *American National Standards Institute Code B31.5* and *API 6d*. Each valve must have the ANSI rating stamped on the body of the valve. Valve selection is made by using the ANSI valve rating chart (see *Table 1*).

Table 1 ANSI Valve Rating Chart

ANSI	MOP (psi)
150	285
300	740
400	990
600	1,480
900	2,220
1,500	3,705

Pipeline designers select valves and components carefully to ensure that the integrity of the pipeline is maintained. A pipeline is like a chain; it is only as strong as its weakest link. A poor selection of valves could greatly reduce the efficiency of the entire pipeline system.

This module provides guidance in the selection of valves to be used on the pipeline. Explanations of proper installation, maintenance, service, and winterizing techniques are also provided. Guidelines for removing a valve from service, servicing the valve, and returning it to service and the proper types of documentation and record keeping are also presented.

2.0.0 ◆ VALVES THAT START AND STOP FLOW

Pipelines use six basic types of valves to start and stop product flow. These types of valves close completely and seal off all flow. These valves are not suitable for throttling flow, as they are designed to be completely open or closed.

2.1.0 Gate Valves

Gate valves (see *Figure 1*) are one of the most common types of valves used on the pipeline. They are the best valves for services that do not require frequent valve operation and for services that require the valve to be either fully open or fully closed. Such valve action is required for pigging operations.

Gate valves should not be used in applications that require throttling of pipeline flow. Close control of flow through a gate valve is almost impossible. Gate valves have very little resistance to flow and exert minimal pressure drop across the valve.

In a gate valve, a gate-like **disc** or slab, actuated by a **stem** and turned by a handwheel, moves up and down across the path of flow and through the valve and **seats** against two seat faces to shut off the flow. Gate valves have four major components: the body, **bonnet**, gate, and stem.

2.1.1 Body

The body of the valve is connected directly to the pipeline and is the main section of the valve. It is the area through which fluid flows. The body of the valve is sized by casting or machining to nominal size so that it corresponds to the size of the pipeline in which it is to be installed. The size of the pipe and the size of the valve body must be the same.

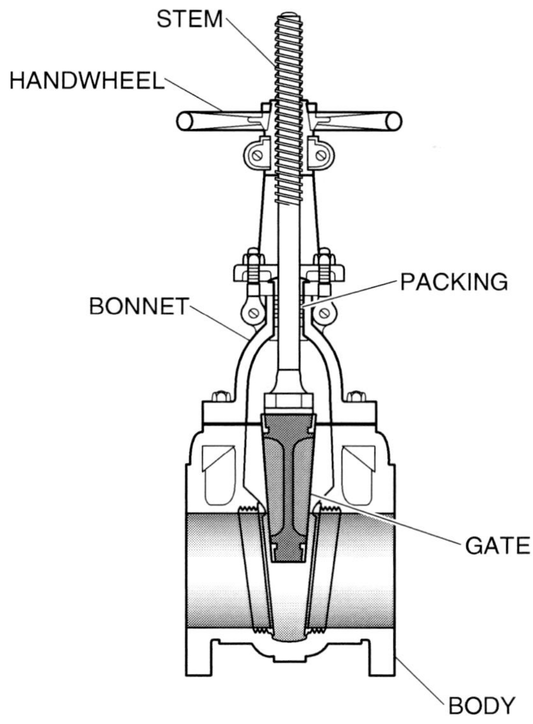

Figure 1 ◆ Gate valve.

2.1.2 Bonnet

The bonnet is located above the body and encloses the **trim**, including the stem, disc, seat ring, disc holder or guide, wedge, and bushings. Various types of bonnets are used with gate valves. In high-pressure applications like a pipeline, most gate valves have a flanged, pressure-sealed bonnet that is bolted to the body using a ring-type gasket (see *Figure 2*).

Valves with bolted bonnets are commonly used for line sizes larger that 1¼ inches. The bolted bonnet gate valve is also the preferred choice in severe

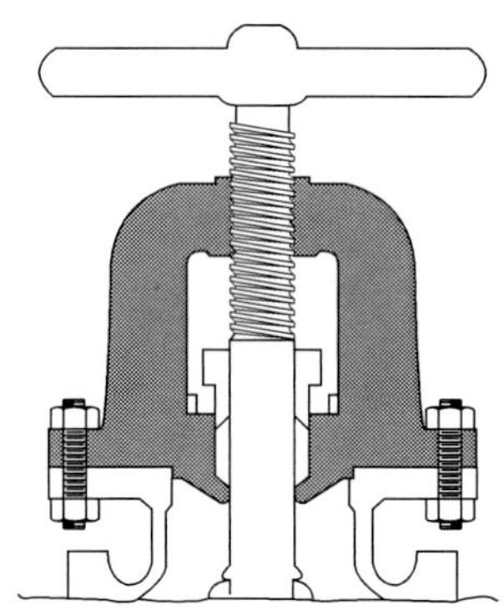

Figure 2 ◆ Flanged, bolted bonnet.

applications where dismantling is frequent. This type of valve must be adjusted after the pipeline is placed in service because application of pressure to the valve relaxes the bolting. This type of valve should be inspected regularly for seal leakage. *Figure 3* shows gate valves with flanged bonnets.

2.1.3 Gate

The gate in a gate valve is the component that actually stops the flow of product through the valve when the valve is closed. **Full port** gate valves provide the maximum internal opening through the valve. This opening will match the internal diameter of the joining pipes. Through-conduit full port valves allow the passage of pigs.

The types of gates used in gate valves include the solid-wedge gate, the double-disc gate, and the simple slab gate. *Figure 4* shows types of gate valve gates.

The solid-wedge is the simplest and most common type of gate. It has good stress resistance and does not trap product in the bonnet when the valve is open. It consists of one piece that is accurately machined on each side to form a leak proof seal with the seat faces.

The double-disc gate consists of two discs that are forced against the seat faces by a spreader located between the discs.

The simple slab gate valve is an efficient, full-port gate valve that uses two floating seats and hardened steel to provide a complete seal with the gate. In corrosive environments, simple slab gate valves with Teflon® inserts and elastomer O-rings are recommended to prolong the life of the valve. As pressure is applied across the port opening, the floating action of the gate and seats ensures a tight seal on both the upstream and downstream sides of the gate.

203F03.EPS

Figure 3 ◆ Gate valves with flanged bonnets.

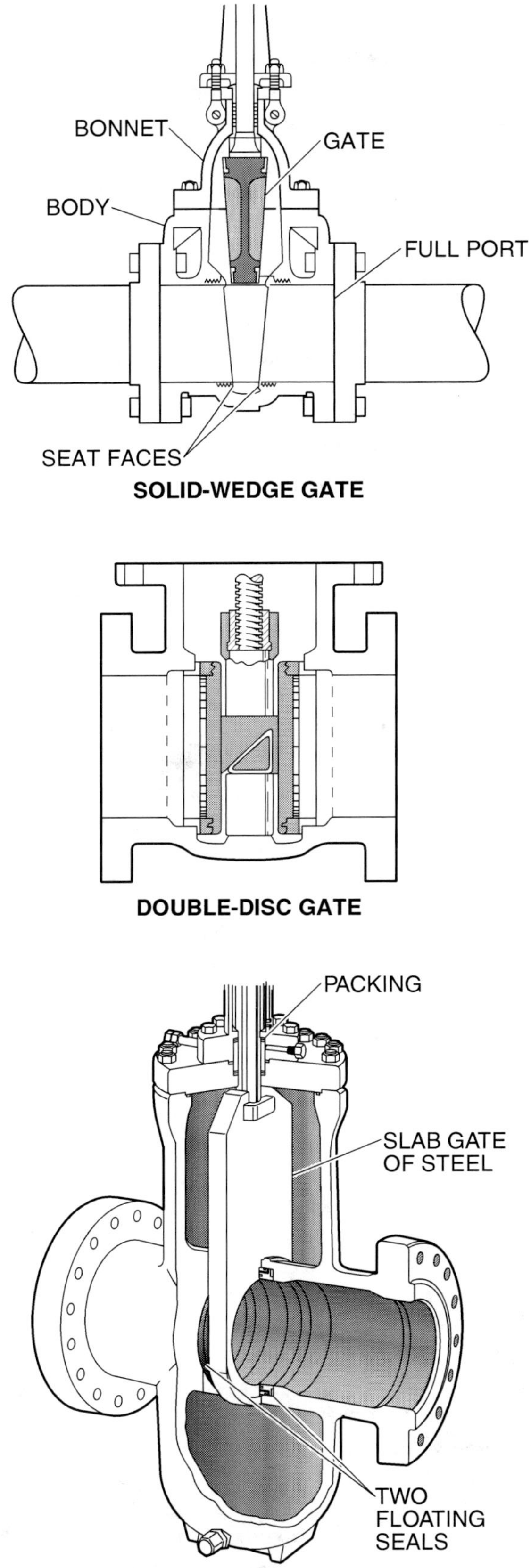

203F04.EPS

Figure 4 ◆ Gate valve gates.

2.1.4 Stem

The valve stem provides a link between the handwheel and the gate. Some handwheels bolt onto the stem. Other handwheels slip over a keyway cut into the stems, and they are secured to the stems by a nut or machine screw. Handwheels vary according to the type of stem they fit. Three general types of stems include the rising stem, the nonrising stem, and the outside screw-and-yoke (OS&Y) stem (see *Figure 5*).

When a rising stem valve is opened, the stem will rise. The height of the stem gives an approximation of how open the valve is. These stems can only be installed where there is sufficient clearance for the handwheel to rise. Rising stems are exposed to the product flowing through the valve and are only used with products that will not harm the threads, such as hydrocarbons, water, and steam.

Nonrising stems are suitable where space is limited. As the name implies, neither the stem nor the handwheel rises when the valve is opened. A spindle inside the valve body is turned when the handwheel is turned, and the gate is raised. Stem wear is held to a minimum. This type of stem is exposed to the product flowing through the valve.

The OS&Y stem is suitable for corrosive products because the stem is not exposed to the product in the pipeline. As the handwheel is turned, the stem moves up through the handwheel. The height of the stem exposed indicates how far the valve is open.

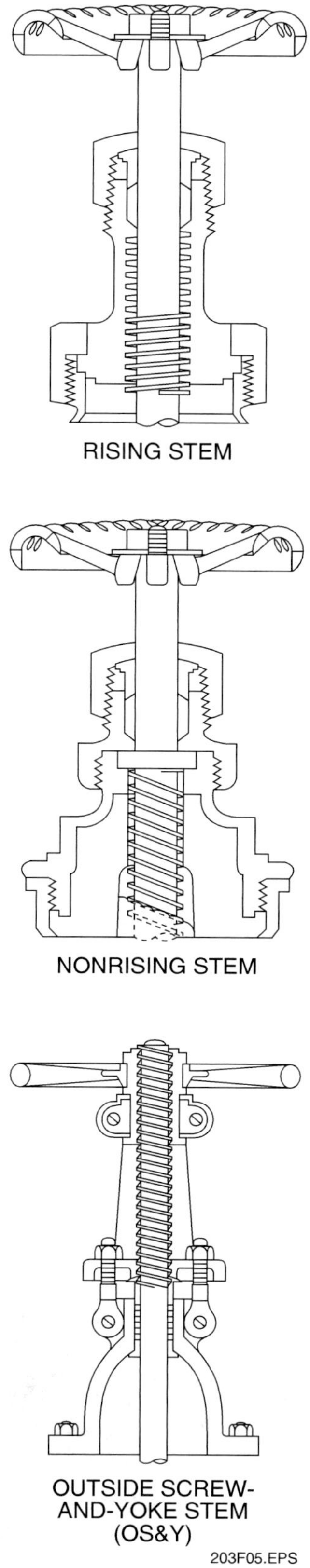

Figure 5 ◆ Gate valve stems.

2.2.0 Ball Valves

A ball valve has a ball with a hole through its center that acts as the control element (see *Figure 6*). This ball is located in the center of the valve body and is attached to a valve stem. A handle is attached to the valve stem. Ball valves turn one-quarter turn to move the hole in line with the flow through the valve. Turning the handle one-quarter turn in the opposite direction moves the solid part of the ball in line with the valve inlet and outlet, stopping the flow through the valve.

Ball valves are quick-opening and easy and inexpensive to repair. The position of the handle indicates whether the valve is opened or closed. If the handle is in line with the flow line of the pipe, the valve is open. If the handle is at a right angle to the flow line of the pipe, the valve is closed. Ball valves are designed to operate either fully open or fully closed.

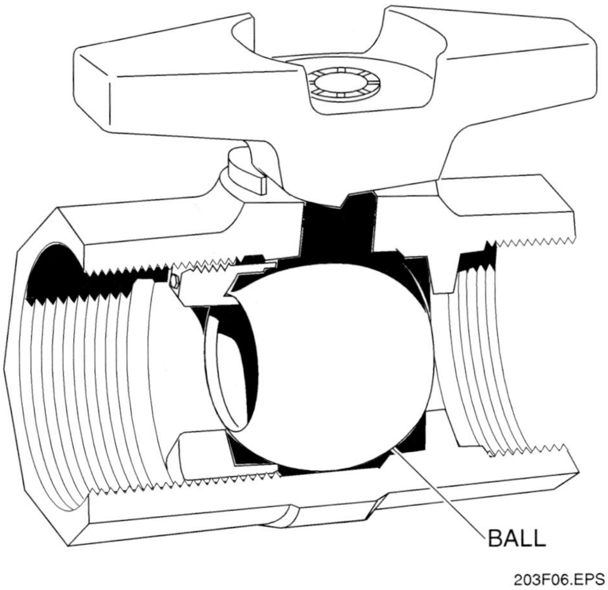

Figure 6 ◆ Ball valve.

2.3.0 Plug Valves

The plug valve, also known as a cock or petcock, consists of a cone-shaped plug that fits into a machined hole. This plug has a hole through it to allow product to flow through the valve when it is in the open position. Plug valves are not used to regulate flow but only to turn the flow completely on or off.

The plug valve can be operated using a handwheel or a wrench attached to the top of the stem. The top of the stem is often a square plug to which a wrench is attached for opening and closing only. The wrench is removed until it is needed again.

As illustrated in *Figure 7*, there are three types of plug valves: regular, short, and venturi. The regular plug valve pattern provides the largest **port** opening available in plug valves. Short-pattern plug valves offer the largest port valve opening available in a size that is consistent with the face-to-face dimensions of a gate valve. The venturi pattern has a smaller port than the other two patterns and provides smooth flow angles into and out of the valve to minimize pressure drop through the valve.

Plug valves are easy to operate and require a minimum amount of installation space. The advantages of a plug valve include the following:

- A fully opened valve exposes no part of the machined surfaces of the plug to erosion or corrosion by the flowing product.
- A fully opened plug valve provides a smooth passageway that does not allow sludge or particles to accumulate.

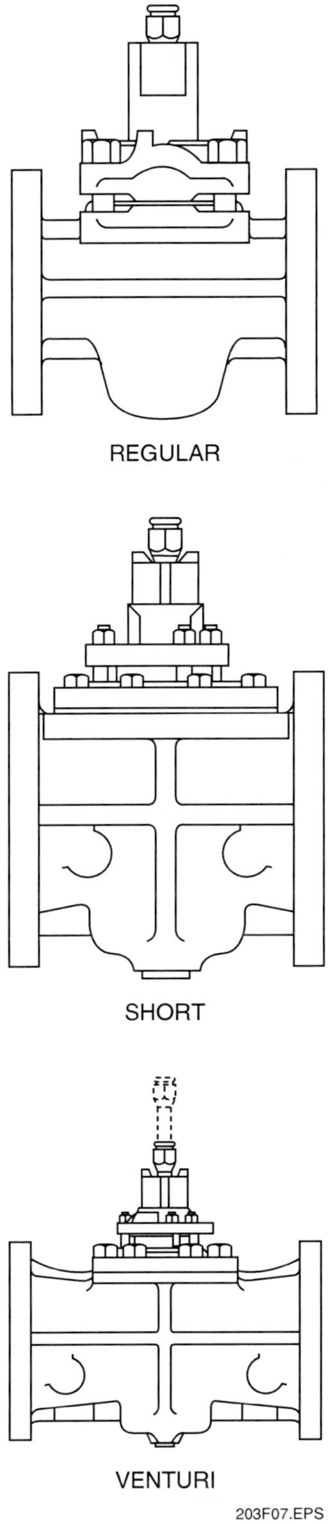

Figure 7 ◆ Plug valves.

- A quarter turn of the wrench or handwheel opens or closes the valve.
- A closed valve exposes only a very small area to the product in the pipeline.

2.3.1 Lubricated Plug Valves

Lubricated plug valves have a lubricating screw at the top of the valve that forces lubricant into grooves in the plug and to the bottom lubricant chamber. These plugs are used in many critical situations that ordinarily would require the plug to be repacked with grease under pressure. There are several disadvantages to using lubricated plugs:

- The lubricant may contaminate the product in the pipeline.
- Lubrication requires extra maintenance.
- The type of lubricant determines the maximum service temperature of the valve.

2.3.2 Nonlubricated Plug Valves

Nonlubricated plug valves have a cam-crank mechanism that lifts the plug and turns it without creating friction between the plug and the seat. These valves are used for service where lubrication is inconvenient or temperature exceeds lubricated plug limits. They are excellent for installations where corrosion requires special alloys and linings. The disadvantages of nonlubricated plug valves are that they cannot be repacked while under pressure and that the seal they provide is not as positive as that provided by a lubricated plug valve.

2.3.3 Three-Way Valves

Three-way valves are multi-port plug valves that are installed at the intersection of three lines. They are used to direct the flow between two of the lines only and to block off the third line. Situations requiring three-way valves include alternating the connections of two supply lines to a common delivery line or diverting a line into one or the other of two possible directions. Three-way valves are designed so that when the plug is turned from one position to another, the channels previously connected are completely closed off before the new channels begin to flow, preventing mixture of fluids or loss of pressure. Three-way valves come in several different arrangements. They can be two- or three-port valves with stops that limit the turning of the plug to two, three, or four positions.

2.3.4 Four-Way Valves

A four-way diverter valve is another variation of the plug valve designed for specialized use and is widely used in the pipeline industry. The four-way valve is central to the operation of bidirectional pipe prover systems. These valves are constructed to very strict requirements.

2.4.0 Expanding Gate Valves

Expanding gate valves (see *Figure 8*) were developed to accomplish **block and bleed** valve operations without the necessity of two gate valves separated by a spool. Originally, the spool was fitted with a bleed valve to guarantee product separation. Expanding gate valves are designated as **double block and bleed** valves because they accommodate the same requirements with a single valve.

The design of expanding gate valves eliminates seal abrasion; it allows the valve to be turned more easily, and when fully closed, a mechanically created, metal-to-metal seal is completed, reinforcing the primary seal. An airtight seal is created both upstream and downstream. Such valves have no reliance on line pressure for a complete seal.

Sealing action is accomplished by two independent seals mounted on a tapered plug. During closing, the mechanical wedge in the plug forces the sealing areas against the machined surfaces of the valve body. The plug and seals do not contact the surfaces until the plug is completely lowered and in position for sealing. After the plug is completely in position, the valve action forces the seals against the valve body, forming complete and separate seals on both sides of the valve. No scrubbing or wear on the seal is possible during this action. The valve may then be bled, and no contamination is possible.

3.0.0 ◆ VALVES THAT REGULATE FLOW AND PRESSURE

Many valves are designed to provide accurate flow and pressure control through a system. The valves used to regulate flow and pressure through a pipeline system include:

- Globe
- Angle
- Y-type
- Butterfly
- Diaphragm

3.1.0 Globe Valves

The most common valve used to control flow is the globe valve (see *Figure 9*). Globe valves are usually installed in lines that are ½ to 4 inches in diameter but are available in sizes up to 12 inches. Globe valves permit close control of flow because the size of the seat-to-globe opening is directly proportional to the number of turns of the handwheel. The number of turns of the handwheel is directly related to the amount of the opening.

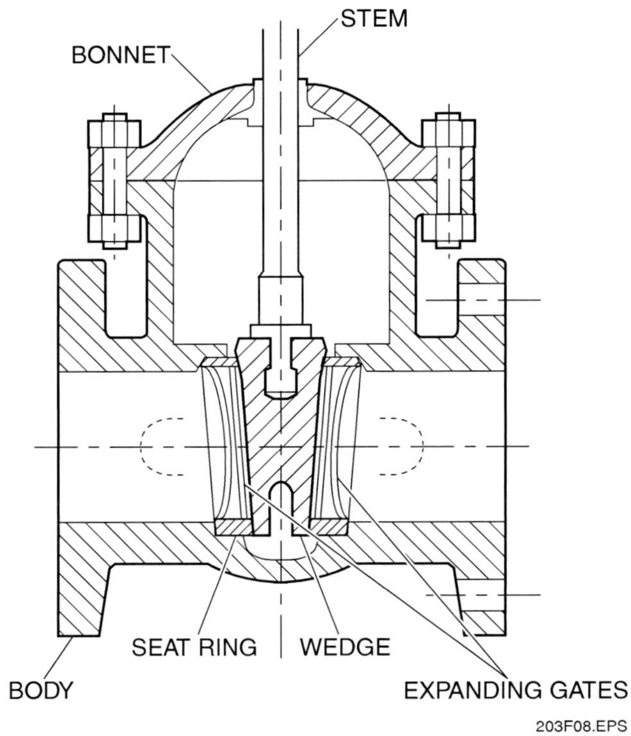

Figure 8 ◆ Expanding gate valve.

The globe valve has a disc that can be partially opened with very little friction to minimize damage to the valve. The design of the globe valve permits flow all around the disc when the valve is partially open, preventing wear or erosion of just one part of the disc. Globe valves are used in pipelines to regulate flow. They are also installed where frequent operation of the valve is required. The major components of a globe valve are the body, bonnet, disc, and stem.

3.1.1 Body

Globe valve bodies have the same general appearance as gate valves. They can be manufactured by the same casting process used for gate valve bodies. They differ from gate valve bodies mainly in seat design. In a globe valve, the seat opening is in line with the valve stem, and the seats are parallel with the line of flow through the valve. Most small valves are built with the seat as a fixed part of the body, and larger, more expensive globe valves are usually made with removable seat rings.

3.1.2 Bonnet

Globe valves are made with several different bonnet arrangements. The more common types of globe valves use a screwed-in bonnet, bolted bonnet, or union bonnet. The screwed-in bonnet should not be used for critical services. If the stem were to become bound to the bonnet by rust, cross threading, or grit in the threads, the bonnet could be accidentally unscrewed while the valve is being opened. Critical service means service in which valve malfunctions could cause the system to shut down or personnel to be injured.

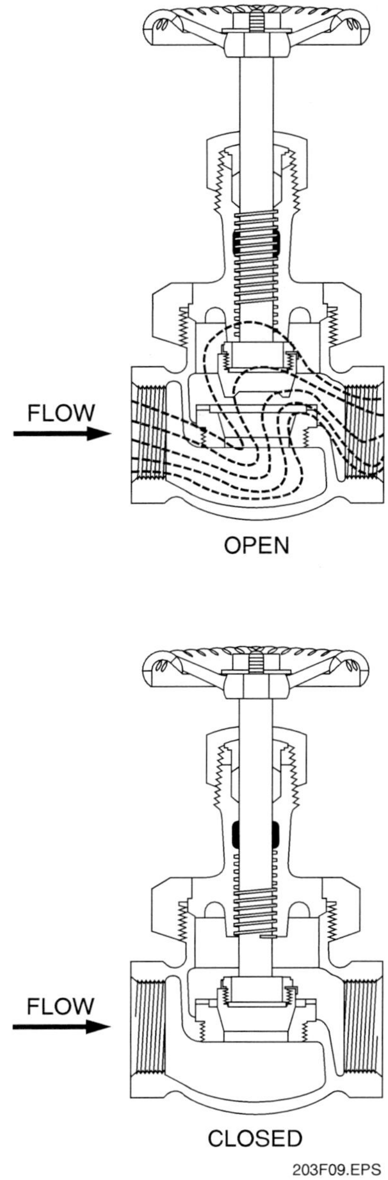

Figure 9 ◆ Globe valve.

3.1.3 Disc

Globe valve discs are made in three designs: the conventional disc, the plug-type disc, and the composition disc (see *Figure 10*). There are also variations within each of these types because each manufacturer uses different design features.

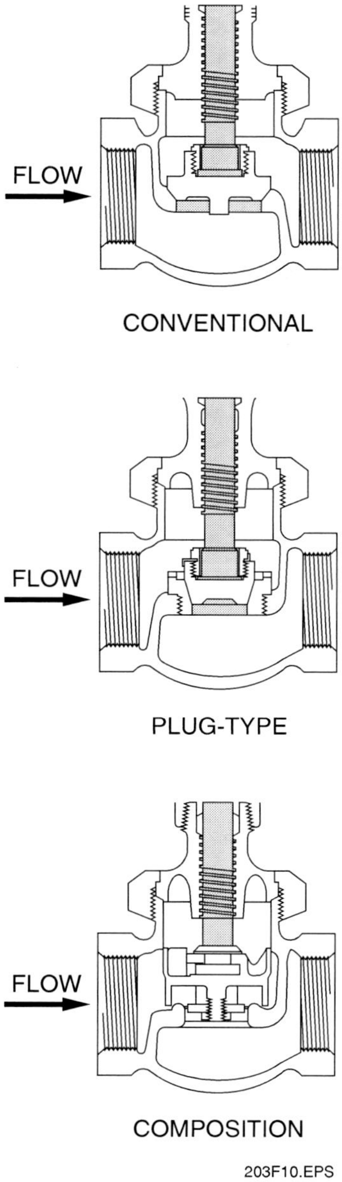

Figure 10 ◆ Globe valve discs.

The conventional disc, which is useful in controlling the flow through a pipeline, has a relatively short taper on its seating face and has a high seating pressure. This enables this type of disc to crush hard deposits that may form on its seating face.

The plug-type disc has a much larger seat contact area than the conventional disc but is otherwise quite similar. Its main advantage over other types of discs is that it offers better flow control of fluid through the valve. The plug-type disc can also withstand more damage from dirt, scale, and solids than the conventional disc before it has to be replaced.

The composition disc consists of three basic parts: a metal disc holder, a disc, and a retaining nut. Composition discs are available in a number of different materials, including rubber, neoprene, and a variety of plastics. They can easily be replaced if they leak during use, but they are not as suitable for throttling products as the other types of discs.

3.1.4 Stem

Globe valves use rising and OS&Y stems. The number of turns of the handwheel determines how far the valve is opened. If 12 complete revolutions of the handwheel are necessary to open the valve all the way, then 3 revolutions will open the valve a quarter of the way, and 6 revolutions will open the valve halfway.

3.2.0 Angle Valves

Angle valves (see *Figure 11*) are variations of globe valves. The outlet of an angle valve is at a 90-degree angle from the inlet. There is less resistance to flow in an angle valve than in a standard globe valve, and the angle valve offers a change of direction in a pipeline without using an elbow. These valves are normally used in non-critical services in place of a globe valve and an elbow. You must be careful when using angle valves because bends in piping systems are subject to stresses that should not be placed on valves.

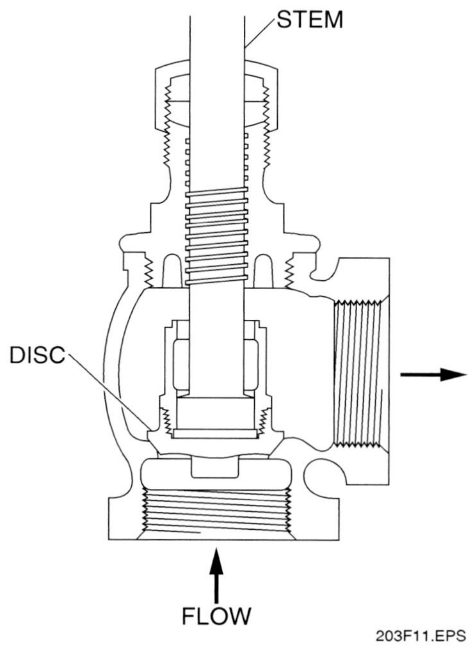

Figure 11 ◆ Angle valve.

3.3.0 Y-Type Valves

A Y-type valve is a cross between a gate valve and a globe valve (see *Figure 12*). It provides straight-through flow with minimum resistance of the gate valve. A Y-type valve provides the throttling ability and flow control of a globe valve. It produces lower pressure drop and turbulence than a standard globe valve and is preferred over a globe valve for use in corrosive services. Y-type valves are also used in many high-pressure applications, including boiler systems. All of the disc and seat designs available in globe valves are also available in Y-type valves.

3.4.0 Butterfly Valves

A butterfly valve has a round disc that fits tightly in a mating seat and rotates 90 degrees in one direction to open and allow fluid to pass through the valve. The butterfly valve can be operated quickly by turning the handwheel or hand lever one-quarter of a turn, or 90 degrees, to open or close the valve.

Butterfly valves can be used completely open, completely closed, or partially open for non-critical throttling applications. Butterfly valves are used in low- to medium-pressure and low- to medium-temperature applications.

When the butterfly valve is equipped with a hand lever, the position of the hand lever indicates whether the valve is opened, closed, or partially open. When the lever is parallel with the flow line through the valve, the valve is open. When the lever is perpendicular to the flow line through the valve, the valve is closed.

Butterfly valves that are 12 inches in diameter or larger are usually equipped with a handwheel or gear-operated **actuator/operator**. This is required because of the large volume of product flowing through the valve and the great amount of pressure pushing against the seat that must be overcome when the valve is being closed.

Butterfly valves have an arrow stamped on the side, indicating the direction of flow through the valve. They must be installed in the proper flow direction, or the seat will not seal and the valve will leak. Three types of butterfly valves include the wafer, wafer lug, and two-flange.

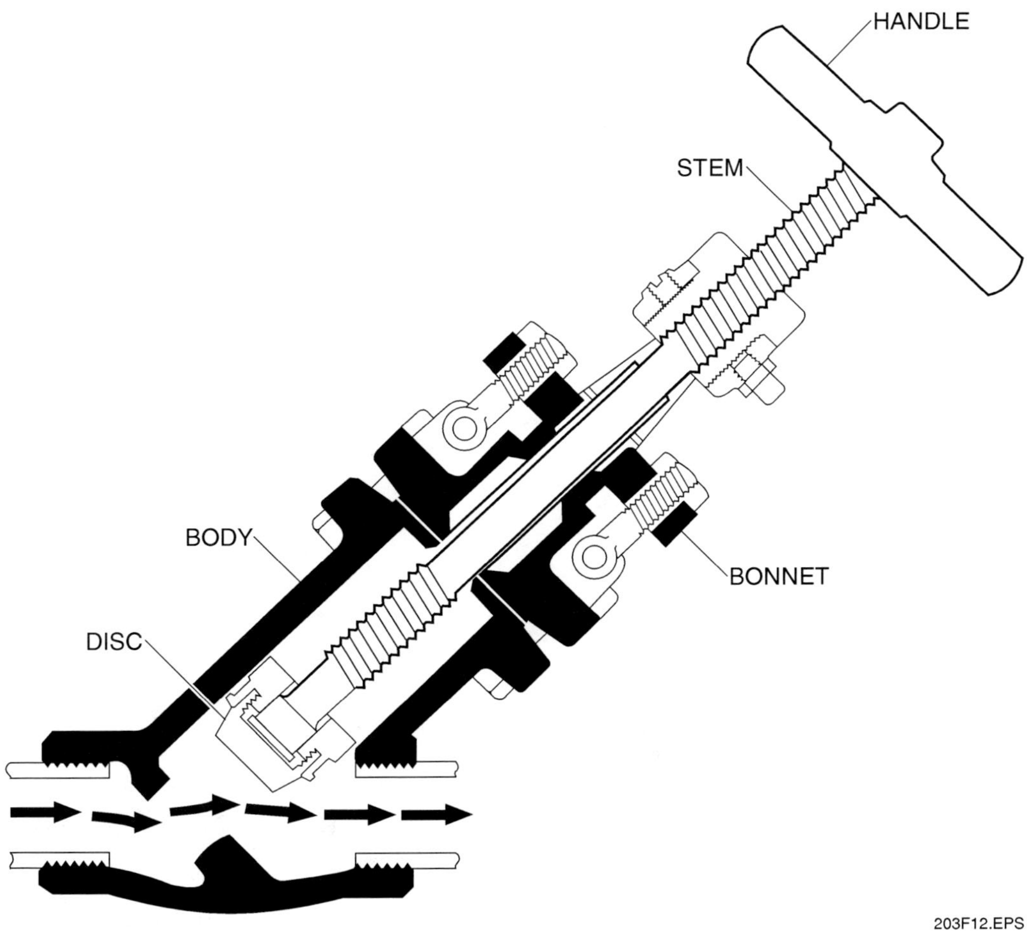

Figure 12 ◈ Y-type valve.

3.4.1 Wafer Valves

The wafer butterfly valve (*Figure 13*) is designed for quick installation between two flanges. The valve seat is lapped over the edges of the body to make contact with the valve faces. Boltholes are provided in large wafer valves to help line up the valve between the flanges. During the installation of a wafer valve, the valve must be in the open position before beginning to apply torque to the bolts in order to prevent damage to the valve seat.

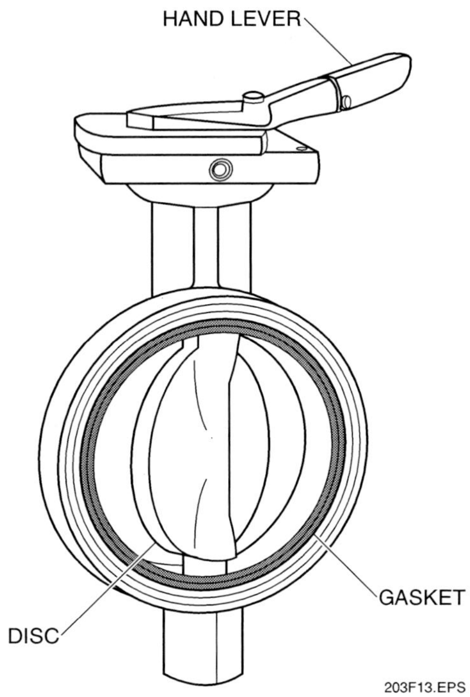

Figure 13 ◆ Wafer butterfly valve.

3.4.2 Wafer Lug Valves

Wafer lug valves (*Figure 14*) are the same as the wafer valves except that they are designed with bolt lugs around the valve body. The lugs on some wafer lug valves are drilled and tapped so that when the valve is closed, a pipeline downstream can be dismantled for cleaning or maintenance while the product and upstream pipeline are left intact. When a tapped wafer lug valve is used for pipeline end applications, only one flange is necessary.

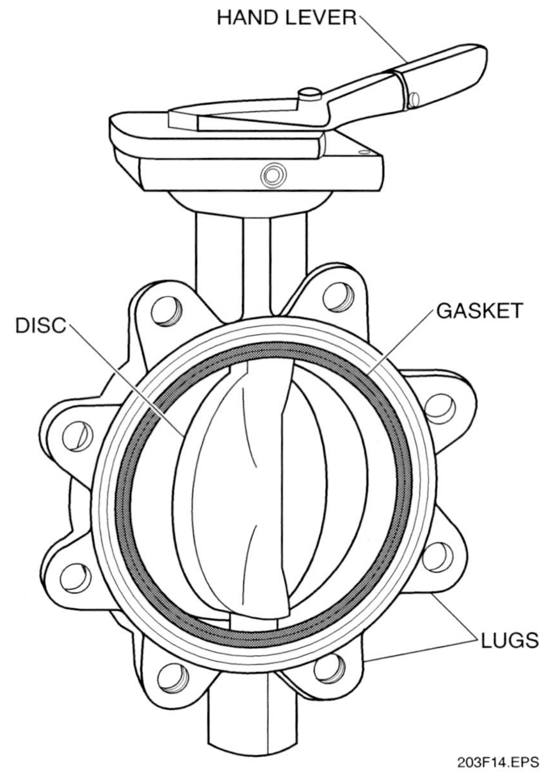

Figure 14 ◆ Wafer lug butterfly valve.

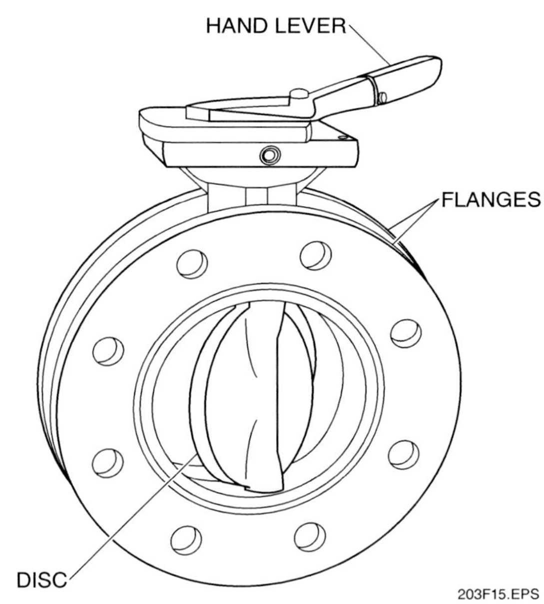

Figure 15 ◆ Two-flange butterfly valve.

3.4.3 Two-Flange Valves

The body of a two-flange butterfly valve (*Figure 15*) is made with a flange on each side. The valve seat is not lapped over the flange ends of the valve, so gaskets are required between the flanged body and the mating flanges of the pipeline. The two-flange valve is made with either machined, flat-faced flanges or raised-faced flanges. The mating flanges on the pipeline must match the valve flanges.

3.5.0 Diaphragm Valves

A diaphragm valve (see *Figure 16*) controls flow by means of a flexible disc that is connected to a compressor by a stud molded into the disc. The valve stem moves the compressor up and down, regulating the flow. Diaphragm valves have no seats because the body of the valve acts as the seat for the flexible disc. The operating mechanism of the valve is not exposed to the product within the pipeline.

These valves can be used fully opened, fully closed, or for throttling flow. They are good for handling slurries, highly corrosive materials, and materials that must be protected from contamination. Many types of fluids that would clog other types of valves will flow through a diaphragm valve.

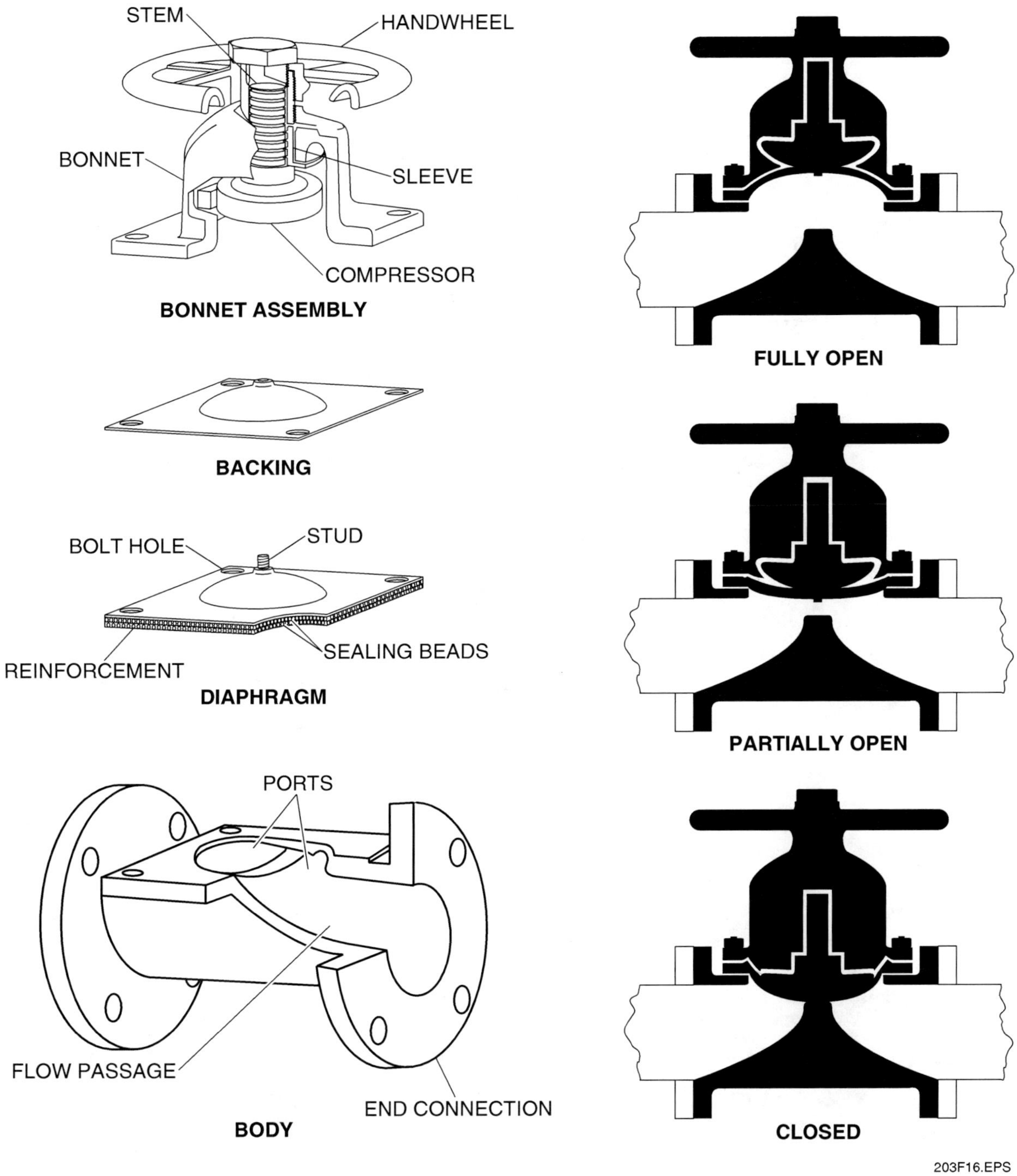

Figure 16 ◆ Diaphragm valve.

3.6.0 Control Valves

Control valves are variations of angle and globe valves that are controlled by pneumatic, electronic, or hydraulic actuators. Control valves normally operate in a partially open position. They are commonly used on the pipeline for pressure and flow control. As explained earlier, control valves are used to control flow, but they also relieve pressure in some applications. Control valves are precision built for increased accuracy in flow control. The control valve is usually smaller than the pipeline to avoid throttling and constant wear of the valve seat (see *Figure 17*).

Control valves are used with pressure-sensing elements that measure the pressure drop at a given point in the pipeline. The flow rate is directly related to the pressure drop. The sensing element sends a pressure signal to a controller that compares the actual pressure drop with the pressure drop for the desired flow rate. If the actual flow is different, the valve controller sends an alarm signal to the control center console. The operator can adjust the valve from the control center through the controller and valve actuator to increase or decrease the flow. *Figure 18* shows a schematic for a control valve in a pipeline.

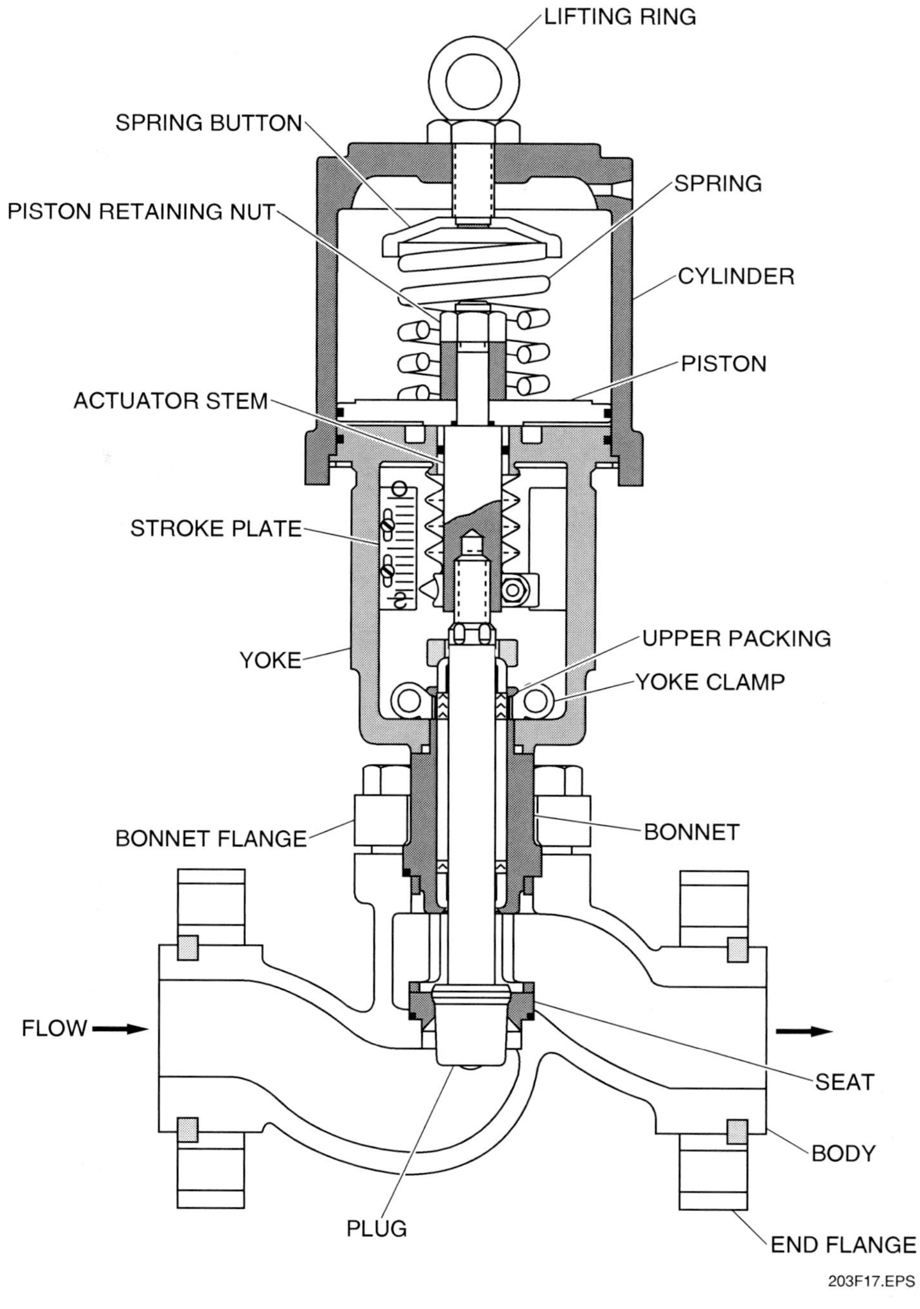

Figure 17 ◆ Control valve.

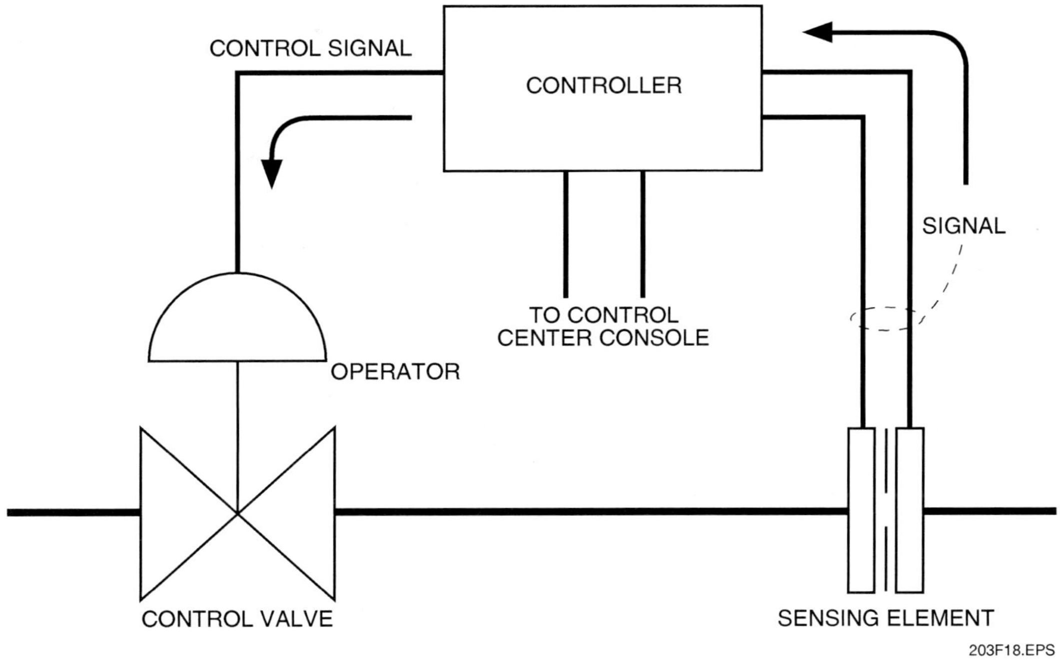

Figure 18 ◆ Schematic for a control valve in a pipeline.

4.0.0 ◆ VALVES THAT RELIEVE PRESSURE

Valves used to relieve pressure in a pipeline, tank, or vessel are known as safety valves or pressure-relief valves. These are installed in pipelines to prevent excessive pressure from rupturing the pipeline and causing an accident or damage to the environment. Safety valves and pressure-relief valves are adjustable and operate automatically.

4.1.0 Spring-Loaded Safety Valves

Spring-loaded safety valves (see *Figure 19*) are used in steam, air, and gas pipelines. Spring-loaded safety valves operate in the closed position until the pressure in the line or unit rises above the preset pressure limit of the valve. At this point, the valve opens fully to relieve the pressure, remains open until the pressure drops, and then snaps shut. Because of this fully open and tight-closing feature, safety valves are commonly referred to as pop-off valves.

Spring-loaded safety valves are most commonly installed in the vertical position. Spring-loaded safety valves should be mounted directly to the tank or vessel being protected, using connecting piping and block valves. Many spring-loaded safety valves are fitted with an outside lever so that pressure can be relieved for maintenance.

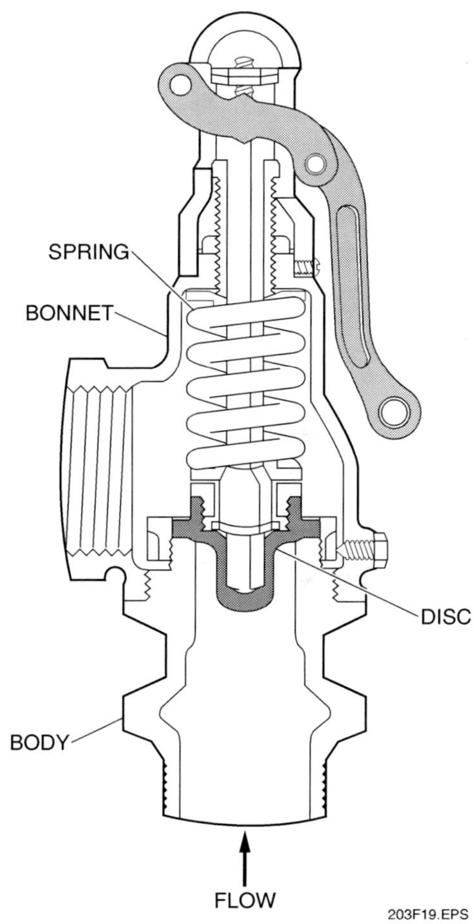

Figure 19 ◆ Spring-loaded safety valve.

Before a spring-loaded safety valve can close, its discharge pressure must drop several pounds below the opening pressure. Spring-loaded safety valves are normally calibrated at the factory and adjusted for appropriate operation.

4.2.0 Pressure-Relief Valves

Pressure-relief valves (see *Figure 20*) are used to relieve pressure on pipelines. Relief valves in small installations are operated by pressure acting directly on the bottom of a disc that is held on its seat by a spring. The amount of spring pressure on the disc is determined at the factory when the valve is manufactured. When internal pressure becomes great enough to overcome the force of the spring on the disc, the disc rises, allowing the liquid to escape. After enough product has escaped to allow the pressure to drop, the system pressure becomes too low to overcome the force of the spring on the disc. The spring then pushes the disc back onto its seat, stopping the flow of product through the valve. The force of the spring is called the compression of the valve.

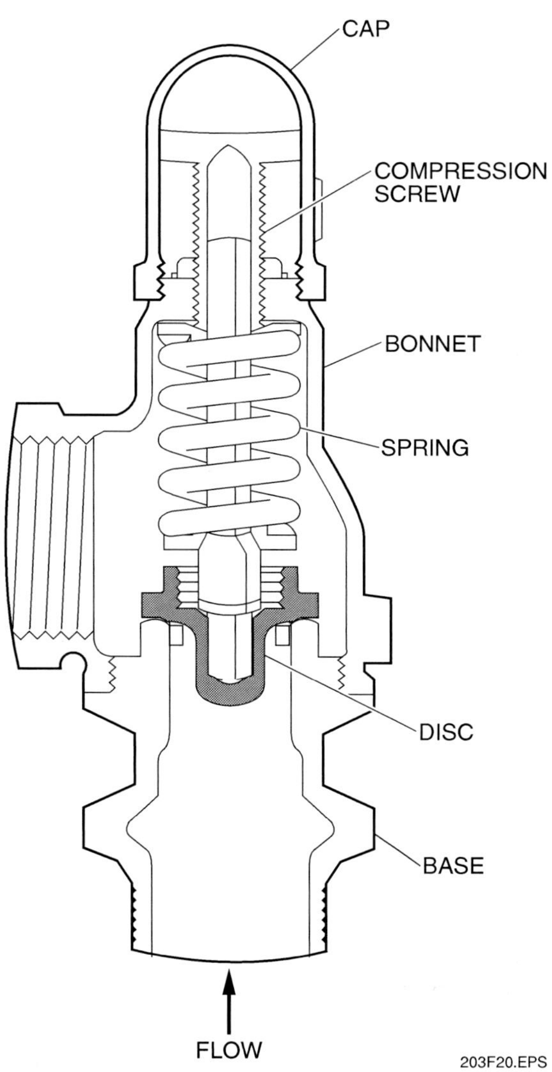

Figure 20 ◆ Pressure-relief valve.

Spring-loaded pressure-relief valves are installed with the stem in the vertical position. The piping connected to these valves must be as large or larger than the valve inlet and outlet openings. A reducer may be used to reduce a larger relief pipe down to the size of the outlet of a relief valve. However, the line size to the inlet of the relief valve cannot be increased above the size of the inlet. Relief piping must be well supported so that line strains will not cause the valve to leak at the seat.

4.3.0 Rupture Disc Devices

Rupture disc devices are pressure-relief devices that are installed to limit the amount of pressure on a line. They are used to protect equipment from overpressure. The pressure-sensitive part of the device is called a rupture disc.

Rupture discs are one-time devices because they operate by rupturing or bursting. When pressure exceeds their design limit and rupture occurs, they must be replaced. Rupture discs are constructed of a very thin metal membrane designed and manufactured to withstand predetermined levels of pressure.

Discs must be selected for the exact application needed. Installation must be in line with expected flow because they are directional in function. Manufacturers now provide many models in quick-change units. If any backpressure is expected, a support member must be included. Rupture discs are used for many manufacturing process installations as well as steam and gas. *Figure 21* shows a rupture disc before and after operation.

5.0.0 ◆ VALVES THAT REGULATE DIRECTION OF FLOW

Valves that regulate the direction of flow are called check valves. Check valves regulate flow by preventing backflow of the product from flowing into the pipeline, permitting flow in only one direction. Check valves are directional valves and must be installed as indicated by an arrow stamped on the side of the valve. A check valve installed in the wrong direction would simply block the pipeline. The force of flow and action of gravity cause the valve to open and close.

Check valves are required to operate in such a way that they do not create extremely high surge pressure when closing. They are also required to open and close smoothly with no erratic behavior of the closing member. Because of differences in operations, even on the same pipeline, valve selection and placement must be carefully planned. For example, when several mainline pumps are on-line

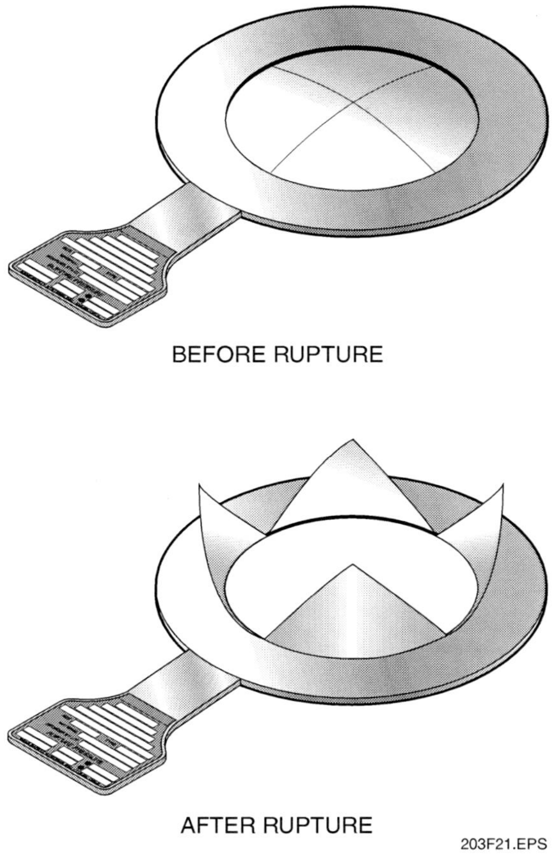

Figure 21 ◆ Rupture disc before and after operation.

and one pump shuts down, the check valve on that pump line must close immediately. On the other hand, if a single main line unit pumping over a long expanse fails, a slow-acting check valve would be sufficient.

Check valves are manufactured in several different designs, including:

- Swing
- Lift
- Piston-controlled
- Ball
- Butterfly
- Foot

5.1.0 Swing Check Valves

The most commonly used type of check valve is the swing check valve (see *Figure 22*). This valve has a disc that swings, or pivots, to open and close the valve. Since check valves provide straight-line flow through the valve body, they offer less resistance to flow than other types of check valves. Swing check valves can be mounted on horizontal installations or vertical installations where product flow is upward.

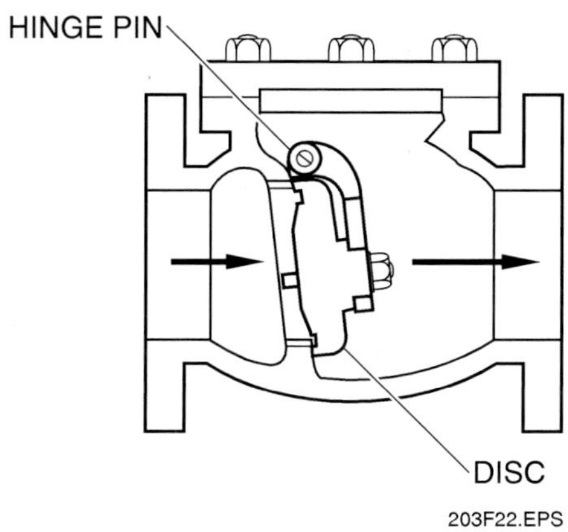

Figure 22 ◆ Swing check valve.

The disc, which is hinged at the top, seats against a machined seat in the tilted wall-bridge opening. The disc swings freely from the fully closed position to the fully open position, which provides unobstructed flow through the valve. The flow through the valve causes the disc to stay open. The amount that the disc opens depends on the volume of product moving through the valve. Gravity or the reversal of flow causes the disc to close, preventing backflow through the valve.

Swing check valves can also be equipped with an outside lever and counterweight to assist the valve in closing. These valves are often referred to as weighted check valves (see *Figure 23*).

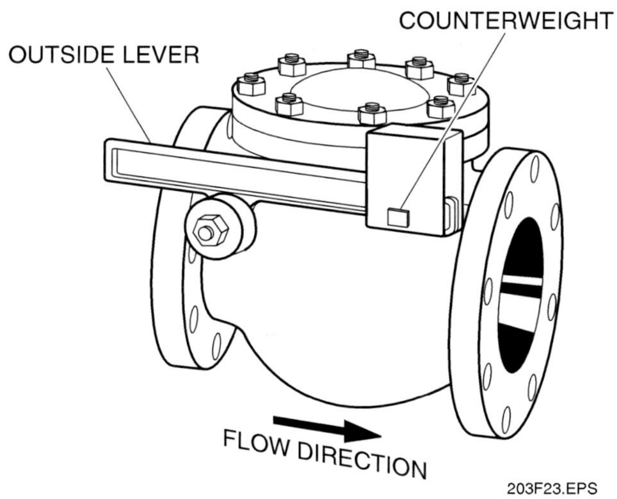

Figure 23 ◆ Weighted check valve.

5.2.0 Lift Check Valves

A lift check valve is designed like a globe valve with an indirect line of flow through the valve. The lift check valve works automatically by line pressure and prevents the reversal of flow through the valve. A lift check valve consists of a disc that is held against the disc seat by gravity or by the pressure of product flowing in the opposite direction through the valve. When product flows through the valve in the correct direction, the force of product lifts the disc away from the seat, allowing product to pass through the valve. Lift check valves can be designed for horizontal or vertical installations (see *Figure 24*) but must only be used for the type of installation for which they are designed.

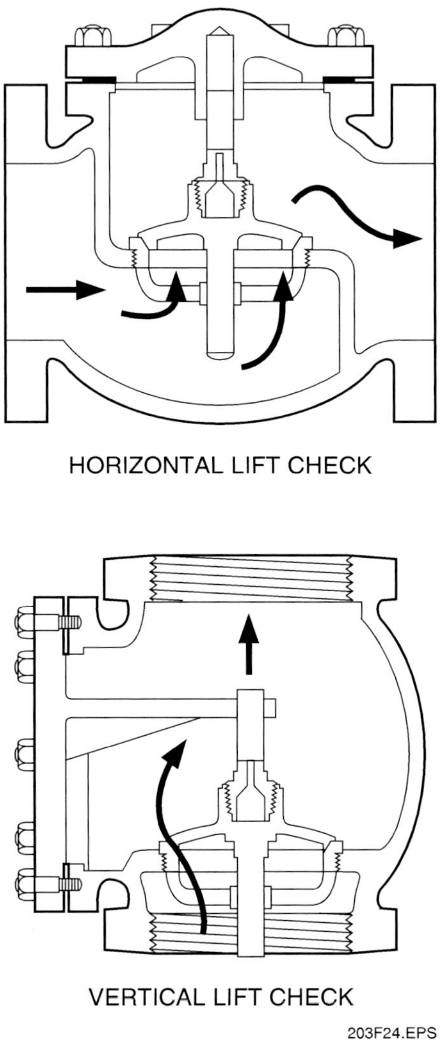

Figure 24 ◆ Lift check valves.

5.3.0 Piston-Controlled Check Valves

Piston-controlled check valves (see *Figure 25*) are designed for the positive prevention of backflow in gas, air, or liquid systems. Flow is directed under the valve piston and lifts it upward. When flow is stopped, the piston descends, closing the valve to prevent flow reversal.

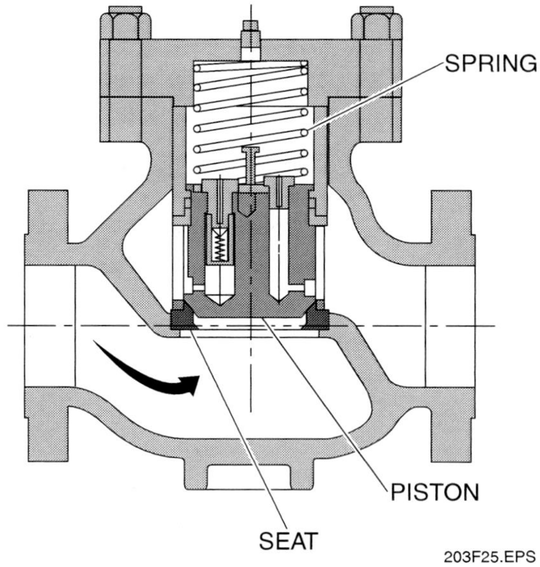

Figure 25 ◆ Piston-controlled check valve.

Pulsating flow through the valve, if unrestricted, would cause the piston to follow the rapidly cycling line pressure, resulting in vibration, chattering, and possible slamming of the piston. The ball check and orifice plug built into the piston are to prevent this from happening.

As the piston rises, the pressure in the Teflon® -sealed cavity above the piston increases, forcing open the ball check. This reduces the cavity pressure and allows the piston to rise smoothly. As line pressure and flow decrease, the piston moves downward. The ball check then closes, and descent of the piston is restricted by fluid flowing through the orifice plug. This equalizes the pressure in the cavity above and below the piston and gives stability to the valve.

5.4.0 Ball Check Valves

Ball check valves operate in the same manner as lift check valves except they use a ball instead of a disc to allow product to pass through the valve. Some ball check valves are designed to work by gravity, while others are spring-loaded. Ball check valves that are not spring-loaded should only be installed in the vertical position. Gravity and the reversal of flow cause the ball to rest against the valve seat, restricting flow through the valve.

Ball check valves are recommended for use in lines in which the pressure changes rapidly because they stop flow reversal more rapidly than other types of check valves. The action of the ball is practically noiseless. The ball rotates during operation and, therefore, tends to wear at a uniform rate over its entire surface. *Figure 26* shows a spring-loaded ball check valve.

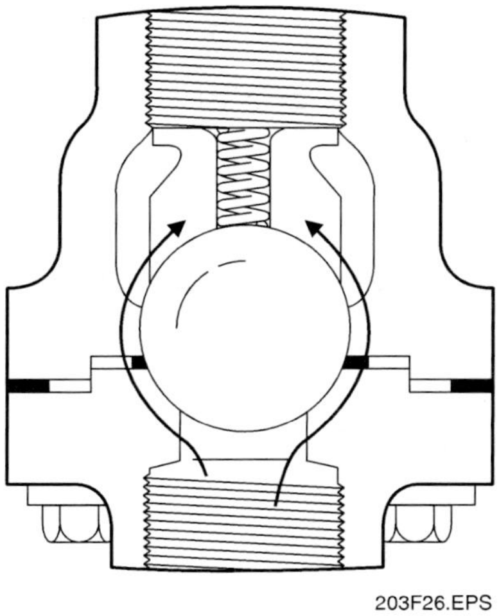

Figure 26 ◆ Spring-loaded ball check valve.

5.5.0 Butterfly Check Valves

A butterfly check valve has two vanes that operate in a fashion resembling the wings of a butterfly (see *Figure 27*). The vanes fold back from a central hinge to open and allow flow but close to prevent backflow. Most butterfly check valves operate completely from the reversal of flow.

5.6.0 Foot Valves

Foot valves (see *Figure 28*) are used at the bottom of the suction line of a pump to maintain the prime of the pump. This type of valve is common in domestic shallow-well applications and industrial applications that do not require a high head. This type of valve operates similarly to the lift check valve. On the inlet side of the foot valve is a strainer that keeps foreign material out of the line. The weight of the product in the pipeline between the pump and the foot valve, sometimes aided by a spring, keeps the valve closed so that the pipeline remains full when the pump is shut down. When the pump starts, the force of the suction pulls the valve open and causes the product being pulled into the valve to push against the outside of the seat to open the valve fully.

6.0.0 ◆ VALVE INSPECTIONS

49 CFR 195.420, *49 CFR 192.745*, and *49 CFR 192.747* require that pipeline companies maintain each valve that is necessary for the safe operation of pipeline systems. Inspections are to be performed at least twice each calendar year on liquid pipeline valves, not exceeding intervals of longer than 7½ months, to make certain that such valves are functioning properly. Inspections on gas pipeline valves must be performed once a year, not to exceed 15 months. Scheduled inspection standards do not usually apply to valves installed only for maintenance purposes, but they are often inspected as a matter of company policy. Inspection is a general procedure that applies to most installations on the pipeline and is carried out much the same anywhere the need is located.

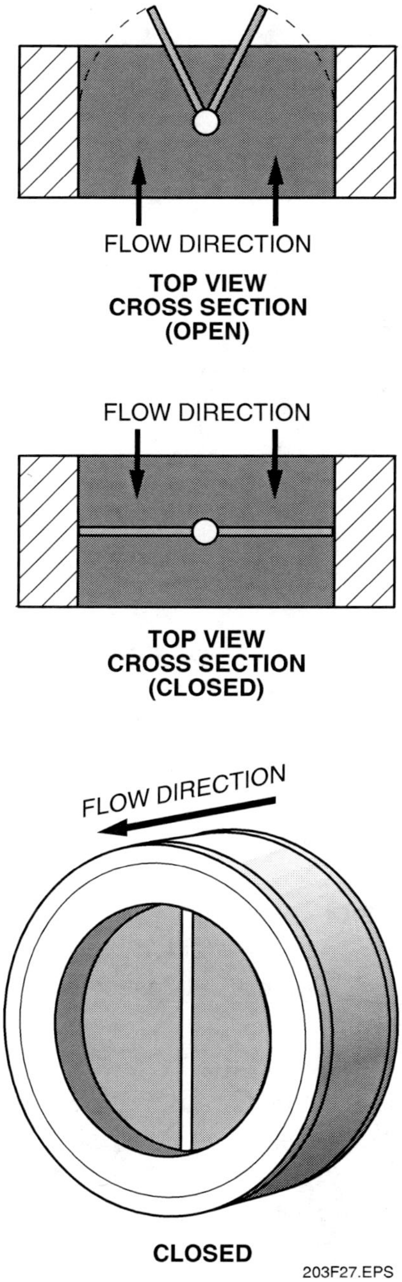

Figure 27 ◆ Butterfly check valve.

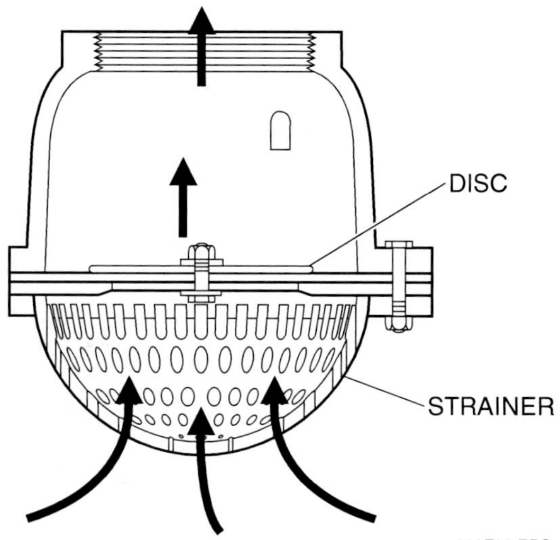

Figure 28 ◆ Foot valve.

Valve inspection is performed to identify any defects that may interfere with valve and pipeline operation. Possible defects may include:

- Leaks
- Coating damage
- Coating failure
- Security problems
- Atmospheric corrosion
- Actuator/operator problems
- Opening/closing problems
- Obstructions

Performing valve inspection involves understanding valve seals and following regulated procedures.

6.1.0 Valve Seals

All valves require a complete seal in four places:

- The valve mechanism itself
- The installation point at the pipe ends
- The installation point of the bonnet
- The valve stem

A stem leak is the first sign of a valve problem. The sooner it is noticed and reported, the more quickly the problem can be solved. Often with stem-type valves, the only action required is to cycle the valve completely by running the stem in and out several times and wiping it off each time. Afterward, observe the valve to see if the leak has stopped. Such a simple procedure might be performed on a smaller yard valve without requiring a call for maintenance. On a mainline valve that might require as much as eight or ten minutes to cycle in one direction, a maintenance operation may be required to have a work order approved by the appropriate departments.

6.2.0 Valve Inspection Procedures

The procedures for inspecting valves consist of the inspection activities performed on each valve that are necessary for the safe operation of a pipeline. This applies to block valves, manifold valves, tank valves, pump/suction valves, check valves, pressure-control valves, and any other valves, except relief valves, that are in service and physically connected to the pipeline. There are four main levels of inspection that can be performed, all of which include these general elements:

- Verifying the location of the valve to be inspected
- Checking the accessibility of the valve
- Verifying the valve number and nameplate data
- Verifying the valve type and manufacturer
- Checking for leaks, damage, and corrosion
- Operating the valve to verify its proper operation
- Returning the valve to its original position
- Relocking the valve in the proper position or providing security as appropriate
- Documenting results

In the case of all four levels of inspection, abnormal operating conditions may be encountered. These may include the leaking of hazardous substances, fire or explosion, activation of a safety device, component failure, or a communications or power interruption. If any such condition cannot be corrected immediately and safely at the site, the control center should be contacted to isolate or shut down the line. The area should be evacuated and defensive action taken according to company procedures.

Anyone performing these inspections must be familiar with company safety procedures, including the use of appropriate personal protective equipment and emergency evacuation and response procedures.

The four main levels of inspection for valves are:

- Routine walk-around
- External integrity
- **Function test**
- Leak test

6.2.1 Routine Walk-Around Inspection

The routine walk-around inspection is made to ensure that no visible leaks or valve damage are present. The steps that may be included in this inspection are:

Step 1 Verify the location of the valve to be inspected to ensure that the proper valve is located prior to operating it, avoiding line upsets.

Step 2 Verify the accessibility of the valve, and remove any obstructions.

Step 3 Verify the valve number and nameplate data. If missing, the nameplate should be replaced per company instructions.

Step 4 Verify the valve type and manufacturer in order to verify documentation.

Step 5 Check for visible leaks to ensure the integrity of all connections and potential leak paths. Check for any signs of other leaks due to maintenance product injection work.

> **WARNING!**
> Upon inspecting a valve, you may encounter an unexpected hazardous liquid, carbon dioxide, or other gas originating from a stem leak or piping connection leak. If the leak cannot be isolated safely at the site, the control center should be contacted to isolate or shut down the line. Evacuate the area, and take defensive action using the proper personal protective equipment.

Step 6 Check for visible damage and indeterminate conditions.

Step 7 Lock the valve in the proper position if applicable. Check the condition of gates, fencing, doors, and/or locks to protect against unauthorized operation or vandalism.

Step 8 Document the results. Documenting data for compliance, historical review, and trending is an important element of valve maintenance.

6.2.2 External Integrity Inspection

The external integrity inspection is a more detailed inspection to ensure pressure boundary integrity, corrosion protection, lack of valve damage, and correct tagging. The steps performed during this inspection may include:

Step 1 Verify the location of the valve to be inspected to ensure that the proper valve is located prior to operating it, avoiding line upsets.

Step 2 Verify the accessibility of the valve, and remove any obstructions.

Step 3 Verify the valve number and nameplate data. If missing, the nameplate should be replaced per company instructions.

Step 4 Verify the valve type and manufacturer in order to verify documentation.

Step 5 Check for leaks to ensure the integrity of all connections and potential leak paths. Check for any signs of other leaks due to maintenance product injection work.

> **WARNING!**
> Upon inspecting a valve, you may encounter an unexpected hazardous liquid, carbon dioxide, or other gas originating from a stem leak or piping connection leak. If the leak cannot be isolated safely at the site, the control center should be contacted to isolate or shut down the line. Evacuate the area, and take defensive action using the proper personal protective equipment.

Step 6 Check for visible damage and indeterminate conditions.

Step 7 Perform a detailed inspection of all pressure boundary connections and ancillary attachments.

Step 8 Check for corrosion, and quantify any existing corrosion to determine the condition of the corrosion protection and coatings. Ensure that the coating is within tolerances for protection from the weather.

Step 9 Verify proper tagging and component identifications to ensure that valve, actuator/operator, and ancillary attachments are correctly identified and accessible.

Step 10 Verify valve, actuator/operator, and other connections.

Step 11 Lock the valve in the proper position if applicable. Check the condition of gates, fencing, doors, and/or locks to protect against unauthorized operation or vandalism.

Step 12 Document the results. Documenting data for compliance, historical review, and trending is an important element of valve maintenance.

6.2.3 Valve Function Test

The valve function test is performed to ensure that the valve is performing its intended function as designed. This may include manually operating the valve or operating it with the use of mechanical assistance, such as an actuator/operator. The result is to ensure that the valve will function correctly upon command. The steps to be performed in this test may include:

Step 1 Verify the location of the valve to be inspected to ensure that the proper valve is located prior to operating it, avoiding line upsets.

Step 2 Verify the accessibility of the valve, and remove any obstructions.

Step 3 Verify the valve number and nameplate data. If missing, the nameplate should be replaced per company instructions.

Step 4 Verify the valve type and manufacturer in order to verify documentation.

NOTE
Locate company procedures for performing a valve function test.

Step 5 Follow the manufacturer's or industry recommendations to prolong valve life.

Step 6 Notify the operations control center and/or affected personnel before beginning the test in order to maintain awareness of activities and to ensure safety.

Step 7 Follow the associated task-specific procedures if applicable. These include procedures to help ensure job safety, task performance, standardization, and documentation of results. They may also include work permit information and notification requirements for the control center, the upstream pump station, and the downstream pump station.

Step 8 Operate the valve if appropriate.

CAUTION
In case of a component failure, including the actuator/operator, you should be prepared to operate the valve manually or with another type of operator.

Step 9 Verify the valve **travel** indication to ensure travel indicators and alarms are properly functioning.

CAUTION
While you are operating a valve, the valve may begin to travel without local input because of a power/communications interruption. If this occurs, immediately contact the control center to verify if the valve travel has been commanded. If the command was not initiated, take steps to stop the valve travel and to reverse it. If the valve travel stopped due to a power loss, you should be prepared to operate the valve either by hand or with another type of operator.

WARNING!
Upon inspecting a valve, you may encounter an unexpected hazardous liquid, carbon dioxide, or other gas originating from a stem leak or piping connection leak. If the leak cannot be isolated safely at the site, the control center should be contacted to isolate or shut down the line. Evacuate the area, and take defensive action using the proper personal protective equipment.

WARNING!
In case of the ignition of a hydrocarbon resulting in a fire or explosion, evacuate the area and notify the fire department.

Step 10 Lock the valve in the proper position if applicable. Check the condition of gates, fencing, doors, and/or locks to protect against unauthorized operation or vandalism.

Step 11 Notify the operations control center and/or affected personnel of test completion to verify the integrity of the system for its return to normal operation.

Step 12 Document the results. Documenting data for compliance, historical review, and trending is an important element of valve maintenance.

6.2.4 Valve Leak Test

The valve leak test is performed to determine the valve's ability to seal and to determine that the **leak-through** or **leak-by** rates are acceptable. Leak-through is a condition in a gate or ball valve in which liquid can leak past both valve seats, causing the valve to leak from the high pressure side to the low pressure side when the valve is closed. For single-seated valves, leak-through is a condition in which liquid leaks by the valve seat, causing the valve to leak internally when closed. Leak-by is an **internal leak** condition in double-seated valves. The steps that may be performed in the leak test are:

Step 1 Verify the location of the valve to be inspected to ensure that the proper valve is located prior to operating it, avoiding line upsets.

Step 2 Verify the accessibility of the valve, removing any obstructions.

Step 3 Verify the valve number and nameplate data. If missing, the nameplate should be replaced per company instructions.

Step 4 Verify the valve type and manufacturer in order to verify documentation.

Step 5 Notify the operations control center and/or affected personnel before beginning the test in order to maintain awareness of activities and to ensure safety.

Step 6 Follow the associated task-specific procedures if applicable. These include procedures to help ensure job safety, task performance, and standardization and to help ensure that results are documented. They may also include work permit information and notification requirements for the control center, the upstream pump station, and the downstream pump station.

Step 7 Follow the manufacturer's or industry recommendations to prolong valve life.

Step 8 Operate the valve if appropriate.

Step 9 Obtain the differential to perform leak testing. Leak rates can be quantified using the known differentials.

Step 10 Leak test the valve. The valve's ability to seal for maintenance, operations, or spill prevention is based on leak-by and/or leak-through rates.

WARNING!
Upon inspecting a valve, you may encounter an unexpected hazardous liquid, carbon dioxide, or other gas originating from a stem leak or piping connection leak. If the leak cannot be isolated safely at the site, the control center should be contacted to isolate or shut down the line. Evacuate the area, and take defensive action using the proper personal protective equipment.

Step 11 Lock the valve in the proper position if applicable. Check the condition of gates, fencing, doors, and/or locks to protect against unauthorized operation or vandalism.

Step 12 Notify the operations control center and/or affected personnel of test completion to verify the integrity of the system for the return to normal operation.

Step 13 Document the results. Documenting data for compliance, historical review, and trending is an important element of valve maintenance.

A number of abnormal operating conditions (AOCs) may be encountered while performing a valve leak test. These AOCs may include the activation of a safety device and failure of a component.

If a valve travels towards the closed position during a function test, the control center should be immediately notified. The system will overpressure if the valve inadvertently closes. If it is verified that this is not an authorized command, actions should be taken to stop the valve travel and to return the valve to its normal position.

If the drain hose connection to the temporary body drain valve fails, immediate steps to close the drain should be taken. The control center should be notified, and the site should be cleaned up as necessary using the proper personal protective equipment.

If the grease fitting fails and product is released, the hose should be put back on the fitting. If this is impossible, the area should be evacuated and the control center should be contacted immediately. If the high-pressure hose fails, the check-in grease fitting should prevent product from escaping.

WARNING!
Keep your hands clear of the hose to prevent a high-pressure injection of grease into your body. You should be prepared to operate the valve either by hand or with another type of operator.

7.0.0 ◆ PERFORMING PREVENTATIVE MAINTENANCE

Every facet of pipeline operation and safety depends on the reliable operation of valves. Valves on the pipeline must be maintained regularly to ensure long life and dependable service. When performing any type of maintenance activities, the valve manufacturer's maintenance procedures must be followed. Removal and reinstallation should be done carefully so the valve will not be damaged. Valves must not be dropped or hit. Valves and materials must never be stored on top of other valves.

Valves should be maintained according to the manufacturer's instructions. Before any work is attempted, the particular valve should be checked to see if the warranty is still in effect. If it is, the work will need to be carried out in an authorized manner.

When performing maintenance, close attention must be given to any valve that requires removal from the pipeline because many valves are directional and must be replaced exactly as they were removed. This especially applies to safety valves, pressure-relief valves, butterfly valves, and others. Such valves should have an arrow on the casting or labels on the ports to designate direction. If the valve is not marked, the technician must determine on which side of the valve the pressure should be located. Gate valves typically require no particular orientation.

Preventative maintenance involves those smaller tasks that, if performed regularly, keep serious problems from occurring as time and service accumulate. Most valves on the pipeline are outside; many are exposed to harsh conditions from weather, dirt, and salt and to corrosion from the elements and products. Most companies maintain a regular schedule of valve maintenance and use a computerized valve maintenance scheduling system that generates timely work orders for valve maintenance so that work can be scheduled. *Figure 29* shows a valve maintenance log with appropriate abbreviations.

Rev. 12/97

Service Performed	Condition of Valve/Site	Deficiency	Action Required	Action Taken
VI - Visual Inspection	AC - Acceptable Condition	DO - Does Not Operate	RP - Replace Part	PR - Part Replaced
MO - Mechanical Operation		LK - Leaking	RR - Repair Required	RP - Repair Performed
LO - Local Mode Operation	NR - Needs Repair / Work	PM - Part Missing		
RO - Remote Mode Operation		DP - Defective Paint	PR - Paint Required	RE - Repainted
RM - Routine Maintenance	RR - Replacement Required	NS - Security	SS - Secure Site	SP - Security Provided
AS - All Services Listed		WO - Wash Out	DW - Dirt Work Required	DP - Dirt Work Performed
	OS - Out of Service	SU - Site Upkeep	SC - Site Cleanup	SM - Site Cleaned Up
		OT - Other (describe)	OT - Other (describe)	OT - Other (describe)

Date	Print Order	Valve Tag No.	Serial Number	Location / Service	Service Performed	Condition of Valve/Site	Deficiency	Action Required	Action Taken
11/01/01	111	42-37B	BR549	#3 Pump Room	VI	NR	DP	PR	RE
11/13/01	112	54-18C	XJ242	Pump Station 14	MO	RR	LK	RP	PR

SAMPLE

REMARKS: ______

DIVISION: ______ STATION: ______ INSPECTED BY: ______

203F29.EPS

Figure 29 ◆ Valve maintenance log.

Some of the procedures that are scheduled when a valve is out of service to maintain efficient operation include:

- Isolation/purging
- Replacing stem O-rings
- Replacing bonnet gaskets
- Repairing/replacing disks

Some of the procedures that are scheduled when a valve is in service to maintain efficient operation include:

- Adjusting packing nut
- Replacing **packing**
- Painting
- Lubricating valves
- Winterizing valves

> **NOTE**
> Before performing any maintenance on a valve in the line, specific company procedures pertaining to work permits, work orders, shutdown procedures, notifications, and documentation must be strictly followed.

7.1.0 General Procedures for Preventative Valve Maintenance

The following procedures are general. Valve manufacturer's guidelines and company procedures vary, as do valve design details. Some companies have specific policies concerning maintenance. Many modern valve designs may allow you to skip steps in the general procedures. Any person attempting to perform any kind of service or maintenance on a valve should first consult company policy and procedure and also refer to the manufacturer's specific instructions for the valve model being considered.

7.1.1 Isolating and Purging a Valve

Valves requiring maintenance must be isolated and purged. This means that the valve must be completely cut off from the system and that the product in the valve and the pipe sections adjoining it must be drained. All pressure must be relieved.

Maintenance work orders must be requested, initiated, and approved by management before any procedure that affects pipeline safety and flow can be performed. The work order is requested by personnel who discover a specific problem or by the regular maintenance program of the company. When a valve requires maintenance, it must be isolated and purged of product before work begins.

Company procedures must always be followed when isolating and purging a valve. Some companies require that the standard operating procedure (SOP) manual be physically on site with the technicians while such operations are being performed. Such procedures often have checklists comprised of steps that must be checked off as each is accomplished. Always refer to the material safety data sheet (MSDS) for the product in the line. Verify that all precautions and specifications being used match those on the MSDS before beginning procedures.

Isolating and purging a valve, in some cases, is considered a major operation on the pipeline. Such operations require changes in operational proceedings, an interruption of flow, and in some events, a shutdown of the pipeline. This must be coordinated through the control center and management. Maintenance records that must be kept for the isolation and purging of a valve may include some or all the following:

- Work order for valve work (including specifications)
- Special work permits
- Control center log
- Excavation work plans

7.1.2 Replacing Stem O-Rings

Some stem valves employ O-rings that create a leak proof seal and hold lubricant around the valve stem. The O-rings are fitted into machined grooves cut around the stem. When an operation is required to seal a leak, the top valve operator must be disassembled and the bonnet removed. Company procedures and the valve manufacturer's specific maintenance procedures must be followed when replacing O-rings in any valve. *Figure 30* shows an O-ring assembly.

7.1.3 Replacing Bonnet Gaskets

Valve designs with bolted or flanged bonnets use gaskets to achieve a complete seal (see *Figure 31*). Gaskets may be of the flat or ring type. The flanged bonnet with a ring-type gasket is the better choice for high temperatures and pressures. When a leak occurs between the bonnet and valve body, the gasket must be replaced.

Some modern innovations and valve designs allow many maintenance operations in the field that previously required shop work. With modern

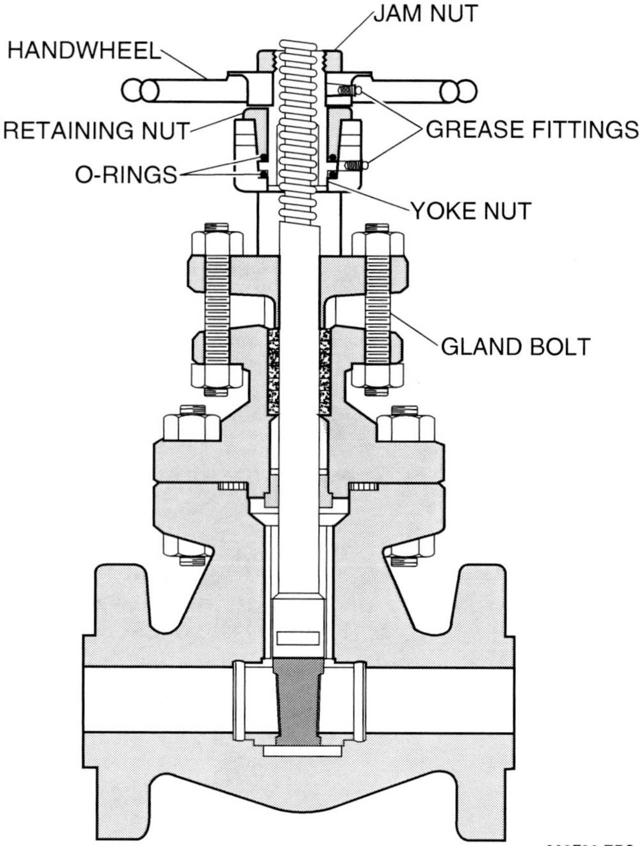

Figure 30 ◆ O-ring assembly.

double gate valves, four-way valves, and others, some formerly major repairs are now done in a few minutes in the field.

7.1.4 Adjusting Packing Nuts

If cycling and wiping the stem several times does not eliminate a valve steam leak, the next step is to tighten the packing nut or gland bolts (see *Figure 32*) as applicable. With passing time, the packing may have dried out and become smaller. This is common in valves that are seldom used. A simple adjustment of the packing nuts or bolts may be all that is required. It is probably better to do so in small increments, and the nuts or bolts should be tightened evenly so as not to damage the existing packing by overtightening. After tightening the nuts or bolts, the valve should be observed to make certain the leak is stopped.

7.1.5 Replacing Packing

A complete seal around the valve stem is difficult to achieve because the stem is required to move. Various types of packing are employed depending on the valve design and product. The packing around the stem prevents the pressurized product

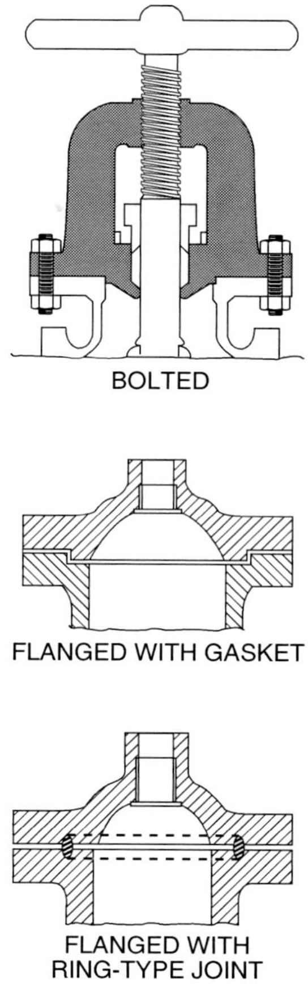

Figure 31 ◆ Bolted and flanged valve bonnets.

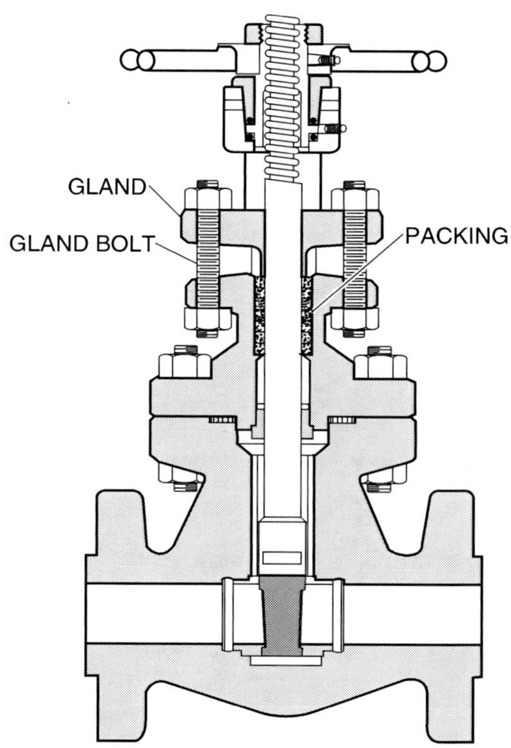

Figure 32 ◆ Gland bolt.

in the pipeline from migrating up the stem. Packing material is compressed into the stuffing box, and a fitting called a gland presses the packing tightly against the stem. The stuffing box requires tightening on a regular basis. Valves that are not used often require tightening before or during periodic use. *Figure 33* shows a valve packing assembly.

Packing is made of solid, braided, or granulated fibers. Cross-sectional shapes include square, round, and wedge. Packing is composed of rubber, asbestos, natural, synthetic, metal, and plastic in a variety of pure and mixed blends. The packing used is determined by the following factors:

- Product in the pipeline
- Maximum operating pressure
- Normal operational temperature encountered
- Minimum temperature encountered
- Metals used in stem construction

In the unexpected event that tightening the valve does not stop the leak, the packing must be inspected and possibly replaced. This procedure is best accomplished in the maintenance shop or other repair area. For large valves where removal from the pipeline would be prohibitive, the standard company policy and procedures must be followed. Repair includes some or all of the following steps:

> **NOTE**
> Placing a valve in a vise requires that some type of protective material be used to cover the vise jaws to protect valve surfaces.

Step 1 Turn the valve to the fully open or fully closed position depending on the manufacturer's instructions.

Step 2 Remove the jam nut from the stem using an appropriate wrench.

Step 3 Remove the operator assembly or handwheel.

Step 4 Remove the retaining nut using an appropriate wrench.

Step 5 Remove the gland bolts from the bonnet.

Step 6 Slide the gland flange off the end of the stem.

Step 7 Slide the packing gland off the end of the stem to reveal the packing.

Step 8 Use a valve packing removal tool (see *Figure 34*) to pull the packing from the gland.

Step 9 Clean all areas of the stem, gland, and stuffing box with an emery cloth.

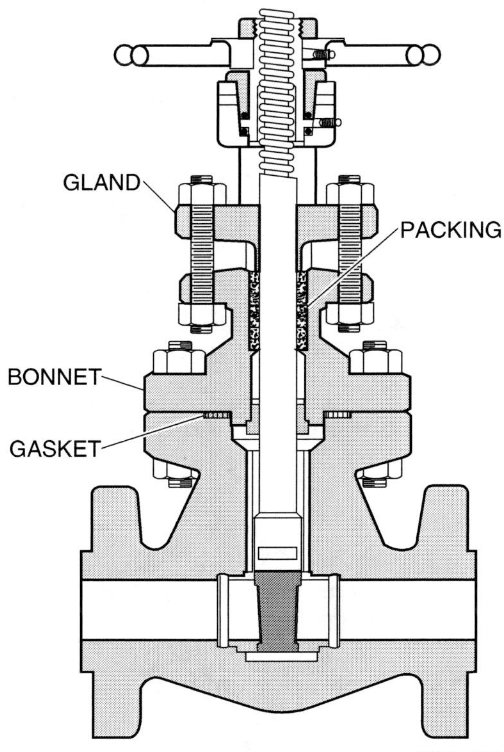

Figure 33 ◆ Valve packing assembly.

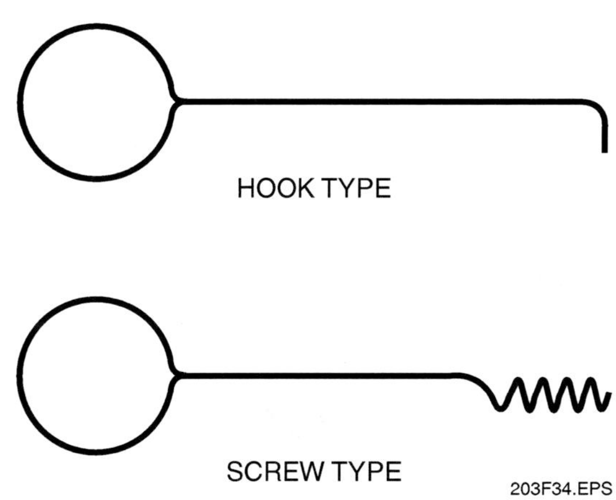

Figure 34 ◆ Valve packing removal tools.

Step 10 Remove all residues of the old packing from the casting and machined areas.

Step 11 Select the manufacturer's recommended packing material.

Step 12 Wrap a strip of the new packing material around the stem.

Step 13 Mark two 45-degree lines where the packing strip overlaps, as shown in *Figure 35*.

Step 14 Remove the packing from stem.

Step 15 Cut at the marked lines with a very sharp knife to obtain a clean cut and fit.

Step 16 Test the cut around the stem to check the fit.

Step 17 Use the first strip as a pattern to cut any others needed.

Step 18 Push the first piece into the stuffing box.

Step 19 Continue adding pieces until the stuffing box is full, staggering the joints (see *Figure 36*).

Step 20 Slide the packing gland onto the stem.

Step 21 Slide the gland flange onto the stem and valve body.

Step 22 Replace the bolts and tighten them.

Step 23 Replace the handwheel.

Step 24 Replace and tighten the handwheel nut.

Step 25 Clean and lubricate the bolt threads.

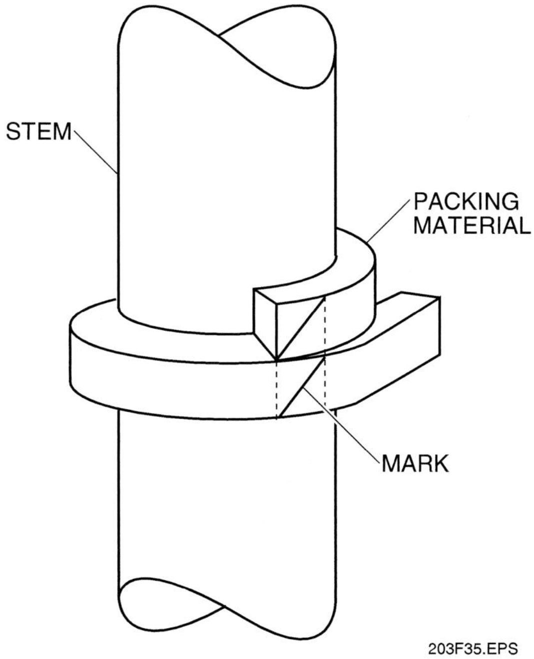

Figure 35 ◆ Measuring valve packing material.

> **NOTE**
> Alternate bolt tightening and perform the final tightening of the bolts with a torque wrench according to the manufacturer's specifications.

Step 26 Reinstall the gland bolts.

Step 27 Oil the stem lightly.

Step 28 Test the operator or handwheel for proper operation.

Step 29 Test the valve as required by company standards.

Step 30 Lubricate all of the grease fittings with a grease gun.

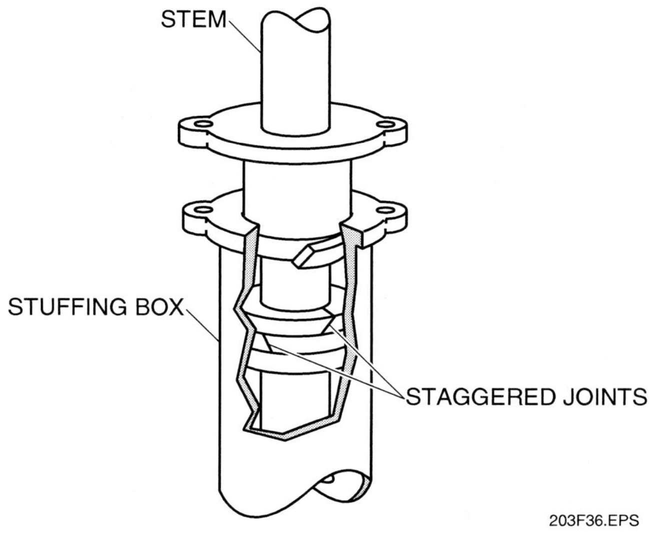

Figure 36 ◆ Staggered packing joints.

7.1.6 Painting

Valves must be painted when new and must be repainted on a regular schedule. Paint is a strong defense against rust and corrosion. Manufacturers coat valves with some type of corrosion-resistant, oxide-type primer, protecting them until delivery. Once valves are delivered, companies paint them according to company policy for resistance to the weather and for product coding (see *Figure 37*). In hot climates, companies paint much of their equipment white. In the colder areas, tanks and equipment are sometimes painted black or some other heat-absorbing color. Some pipelines signify products flowing by color-coding the pipe with paint. *Figure 38* shows two color-coded pipelines.

During inspections, close attention must be paid to the condition of the paint and coatings. Damaged paint allows rust and corrosion to begin.

203F37.EPS

Figure 37 ◆ Contrasting paint scheme.

203F38.EPS

Figure 38 ◆ Two color-coded pipelines.

If there is a chip, ding, or flaked area, it must be touched up before rust and corrosion advance. The operator must arrange to have it done as soon as possible.

7.1.7 Lubricating Valves

Objects that move produce friction and wear. Equipment exposed to the elements tends to experience the drying out, hardening, and deterioration of important components as time passes. Dirt and grime collect in unacceptable places and cause problems with movement, increasing wear, corrosion, and damage to machined and polished surfaces. Periodic lubrication of certain parts of a valve prevents or at least slows down the wearing process, greatly extending the life and efficiency of valve components.

Most periodic lubrication is accomplished in several simple procedures. However, there are other serious considerations that must be included in the work:

- Climate
- Exposure of the valve
- Application of the valve
- Product to be used within the valve

For example, some types of grease become very stiff as the temperature drops. If a valve were outside in a cold climate, a completely different type of grease would be required than that required for a valve exposed to a warm environment. Always use the same grease type as what is currently in the valve. The manufacturer's and company's instructions should be consulted first.

CAUTION

If the wrong type of grease is used, you may damage the operation of the valve.

Some valves require lubrication from a simple grease gun. Others are constructed with a factory-designed lubrication system that only needs a bolt or screw to be turned to force lubrication from an inner chamber to the appropriate locations.

Some lubricants do not endure moisture or rain as well as others. If the valve is exposed to rain, snow, ice, or immersion, moisture may destroy the lubricant if the wrong one is chosen. In areas prone to flooding, tidal activities, or harsh environmental influences, such as blowing sand or dust or severe temperatures, special lubricants will often be required.

Operators and technicians must become completely familiar with the types of valves being used and the types of maintenance products required for their job assignments. They must also become familiar with the types of lubrication systems each valve type employs.

Applying grease with a brush and/or grease gun is a common procedure performed on stem-type valves on the pipeline. This procedure is performed on a variety of stem-type valves that are accessible to personnel. Keep in mind that a grease gun uses hydraulic pressure to move grease from the pump into the fitting. In cold weather, greater pressure may be required to apply grease to the valve. Pumping may require extra effort. *Figure 39* shows a stem requiring lubrication with a brush.

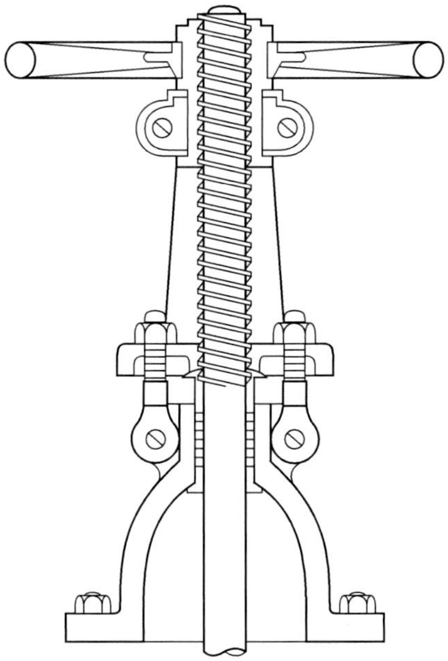

Figure 39 ◆ Stem requiring lubrication with a brush.

Steps to lubricate a stem valve may include:

Step 1 Turn the valve to the completely open position so that the stem threads extend.

Step 2 Clean the threads of all rust, dirt, and grime with a wire brush, solvent, and rags.

Step 3 Wipe off all residues from the moving parts and seal areas with rags.

Step 4 Locate and remove the safety caps from the grease fittings on the sides of the valve body.

NOTE

For those valves that do not have grease fittings, apply the recommended grease heavily to the outer threads of the stem with a paintbrush, and cycle the valve several times.

Step 5 Pump the grease gun several strokes, and operate the valve an inch in each direction to spread the grease on the seals.

Step 6 Pump the grease gun lever several more times on each fitting. *Figure 40* shows a valve with grease fittings.

Step 7 Operate the valve completely several times to spread the lubrication.

Step 8 Move the valve to the fully open position, and pump the grease gun several more strokes to fill the cavity.

Step 9 Remove the handwheel nut and handwheel, exposing the bearing housing.

Step 10 Remove the setscrew holding the bearing assembly in place.

Step 11 Remove the individual bearings and races from the housing.

Step 12 Clean and inspect the handwheel bearings and races.

Step 13 Handpack the handwheel bearings and races.

Step 14 Reassemble the races and bearings into the housing.

Step 15 Slide the bearing housing back over the stem.

Step 16 Tighten the assembly housing setscrew.

Step 17 Reinstall the handwheel and nut.

Step 18 Torque the handwheel and nut according to the manufacturer's specification.

Figure 41 shows a typical operator bearing assembly.

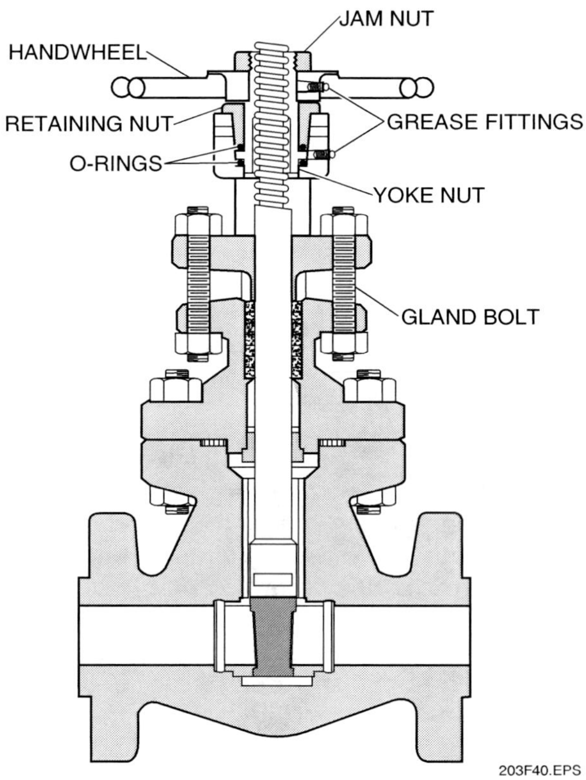

Figure 40 ◆ Valve with grease fittings.

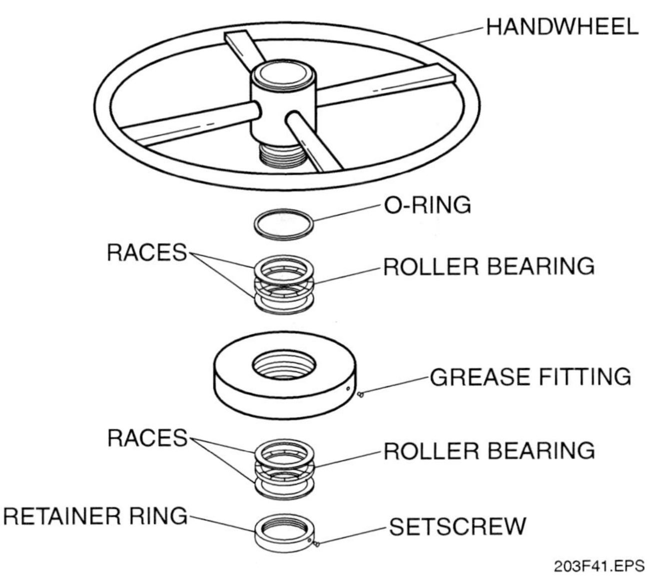

Figure 41 ◆ Operator bearing assembly.

7.1.8 Maintaining Lubricated Valves

Lubricated valves have an inner lubricant supply or cartridge within the body of the valve. At specified intervals, maintenance technicians need only to turn the lubricating screw. Lubricant is forced into grooves in the plug, to the bottom lubricant chamber, and to other mechanisms as required. These valves are especially important to critical operations that require the plug to be repacked with lubricant under pressure. The negative aspects of lubricated valves are:

- Product in the pipeline may become contaminated.
- Extra maintenance may be required.
- The lubricant system may determine the useful life of the valve.
- The lubrication intervals must be effectively tracked.

Figure 42 shows a lubricated, tapered plug valve.

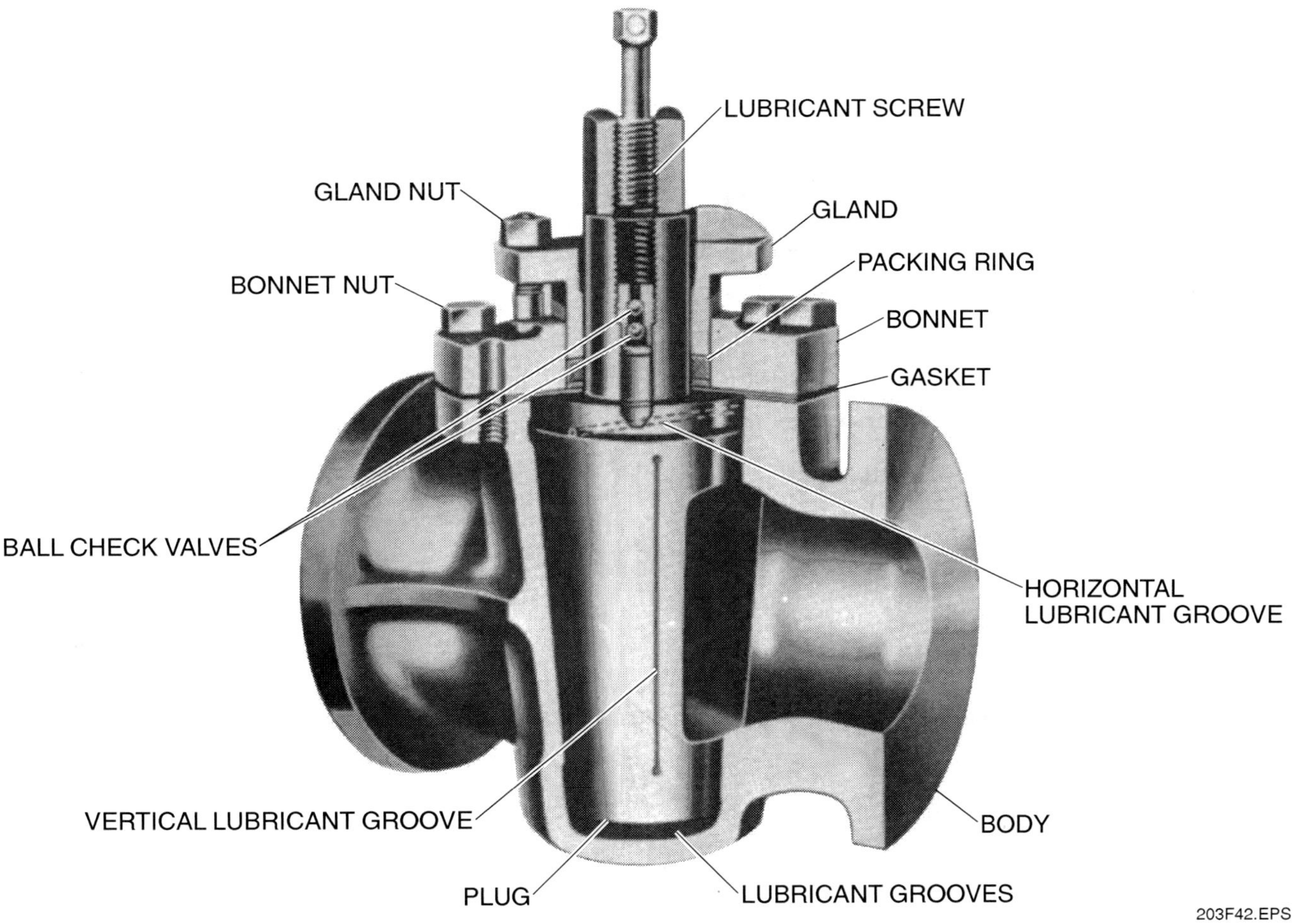

Figure 42 ◆ Lubricated, tapered plug valve.

7.1.9 Winterizing Valves (in Service)

Winterizing valves is an extremely important part of maintenance on the pipeline if the temperature is expected to drop below freezing, even for a short time. Winter temperatures present serious hazards on the pipeline. When the temperature drops below 32°F, equipment that is exposed can experience serious and destructive damage from ice forming inside the unit. Valves are very susceptible to damage and are far more expensive to repair or replace than sections of pipe. If a valve is allowed to freeze and rupture, not only is equipment and production lost, but there is danger to personnel and to the environment. The damage may not be obvious at first sight because the leak will not be revealed until the valve has thawed.

The solution to this problem is to prepare valves for cold weather before it occurs. After the temperature is below 32°F, any effort might prove useless because damage may have already occurred. Even if the casting of the valve is not ruptured, serious damage may have been done to the mechanical parts and seals inside. Certain maintenance procedures will eliminate these occurrences, saving much damage, time, and expense.

Larger valves have cavities in which moisture collects. This moisture may come from wet product, residues of water left in the line after an operation, or condensation. Moisture must be eliminated from the inside of the valve before cold temperatures are experienced. Technicians must check every valve. Valves prone to moisture collection have **drain and vent plugs** located in lower parts of the castings so that they may be drained (see *Figure 43*).

In some locations, draining valve moisture may be a daily task. Company policy must always be followed. Because opening the drain plug will release water contaminated with product, use a receiving vessel to catch the drainage so that it can be properly disposed of. If you are draining a large in-service valve, the pipeline may have to be shut down for a short time to relieve pressure on the valve.

Follow these steps to winterize a valve:

Step 1 Lock out and tag all sources of energy associated with the valve.

Step 2 Place a receptacle under the drain plug or valve to be opened.

Step 3 Close the valve completely.

Figure 43 ◆ Valve with a bottom drain plug.

Step 4 Isolate the valve from the pipeline flow, and drain down the pressure.

Step 5 Open the bottom drain plugs or valves as appropriate to release the moisture.

Step 6 Close the drain plugs or valves after all the moisture has drained from the valve.

> **NOTE**
> Some companies will use antifreeze.

Step 7 Open the pipeline valves, and return them to service.

Step 8 Remove all lock out devices and tags.

Step 9 Remove the receiving vessel, and empty it into a sump or other appropriate holding vessel.

8.0.0 ◆ TERMS AND DEFINITIONS

The following list highlights terms that are commonly used when performing valve inspections. It is important to be familiar with these terms and their definitions:

- *Body bleed* – Opening a body drain or vent to bleed off, or reduce, the internal body pressure on double-seated valves in either the fully open or closed position.
- *External leak* – A leak resulting in the escape or release of product from a valve into the environment.
- *Good working order* – Required by DOT regulations and interpreted to mean that a valve's performance, based on in-service criteria, meets all of the necessary functions of the valve.
- *Hand clutch* – A mechanical means of disengaging the motor drive and engaging the handwheel.
- *Sealant injection system* – Obtains tight shut-off in an emergency situation. A sealant is injected into a specially designed groove in the seat rings. This is available for most ball valves and gate valves.

Summary

Pipelines use valves to start, stop, and regulate flow, relieve pressure, and prevent reversal of flow. The type of valve used is determined by the type of application. Valves are chosen based on the nature of the product in the pipeline, the temperature and pressure of the system, and the desired operation of the system. Valves are manufactured from various materials that are selected to offer the best service for the product being handled.

Representing a sizeable portion of pipeline investment, valves must be cared for and maintained so that their service life and efficiency is preserved as long as possible. Being required by law in many cases, valve maintenance is one of the most important tasks on the pipeline because it ensures the safety, efficiency, and economics of operation.

Maintenance begins with receiving, handling, and properly storing each valve until its installation. Once the valves are installed, proper care ensures that they operate efficiently and safely through all conditions. Special maintenance procedures to prepare for winter ensure safe valve operation in cold weather.

Once the valve is put in place, care exercised in handling and installation pays large dividends in the service life and dependability of the unit. Proper support of the valve, seating of gaskets, torquing of bolts, and care in following manufacturer's instructions all work together to provide extended leak-free operation, ensuring the continued safety and efficiency of the pipeline.

Review Questions

1. All valves have _____ common characteristics.

a. two
b. three
c. four
d. five

2. The type of gate valve that provides the maximum internal opening is a(n) _____ gate valve.

a. expanding
b. full port
c. venturi
d. open center

3. A ball valve has a ball with a(an) _____ in the center.

a. groove
b. axle
c. depression
d. hole

4. A plug valve consists of a _____-shaped plug that fits into a machined hole.

a. box
b. cone
c. ball
d. cylinder

5. Sealing action on an expanding gate valve is provided by _____ independent seals.

a. two
b. three
c. four
d. five

6. Globe valves are available in sizes of ½ to _____ inches in diameter.

a. 2
b. 6
c. 8
d. 12

7. Screwed-in bonnet globe valves should not be used for _____ services.
   a. petroleum
   b. high-pressure
   c. low-pressure
   d. critical

8. Angle valves are variations of _____ valves.
   a. globe
   b. check
   c. gate
   d. butterfly

9. Butterfly valves rotate _____ degrees to open.
   a. 45
   b. 60
   c. 90
   d. 180

10. A wafer valve must be _____ during installation.
   a. closed
   b. open
   c. lapped
   d. calibrated

11. Diaphragm valves have no _____.
   a. bolts
   b. flanges
   c. ports
   d. seats

12. Control valves are variations of _____ valves.
   a. angle and globe
   b. wafer and diaphragm
   c. gate and plug
   d. check and ball

13. Valves used to relieve pressure in a pipeline are known as _____ valves.
   a. control
   b. flow
   c. safety
   d. operation

14. Rupture disc devices are _____ devices.
   a. control
   b. pressure-relief
   c. shutoff
   d. directional

15. Valves that regulate the direction of flow are called _____ valves.
   a. check
   b. gate
   c. relief
   d. rupture

16. The most common type of check valve is the _____ check valve.
   a. gate
   b. swing
   c. reverse
   d. flow

17. A butterfly check valve has _____ vanes.
   a. two
   b. three
   c. four
   d. five

18. Inspections on liquid pipeline valves must be performed _____.
   a. once every two years
   b. once a month
   c. twice a year
   d. once a year

19. The first sign of a valve problem is a _____.
   a. crack
   b. torn seal
   c. gate hole
   d. stem leak

20. Some stem valves use _____ to create a leak-proof seal.
   a. rubber
   b. O-rings
   c. gates
   d. bearings

GLOSSARY

Trade Terms Introduced in This Module

Actuator/operator: A device that converts manual, hydraulic, pneumatic, or electrical energy into mechanical motion to open and close a valve.

Block and bleed: The capability to obtain a seal across the upstream and downstream seat seals of a valve when the body pressure is bled off to the atmosphere through blow-down valves or vent plugs. It is useful in testing the integrity of seat seals and in accomplishing minor repairs under pressure.

Body: The principal pressure-containing part of a valve in which the disc, seat, and valve ports are located. It is connected directly to the pipe by threaded or flanged connections.

Bonnet: The part of a valve that is located above the valve body and contains the trim.

Disc: The moving part of a valve that directly controls flow through that valve.

Double block and bleed: A valving arrangement or valve type, such as a general twin seal or grove tru-seal, that ensures no flow in a line even though the valve may leak. It consists of two block valves in the main line with a small bleeder valve draining the line between the block valves.

Drain and vent plug: A special pipe plug having a small vent valve that is operated with an allen wrench. These special plugs are located at the bottom of most ball valves. The plug and valve are used to vent body cavity pressure and, when removed, to blow out debris or flush the cavity.

Full port: The maximum internal opening for flow through a valve that matches the inside diameter of the pipe used.

Function test: The operation of a valve to assure that it is performing its intended function as designed. This may include manually operating the valve or using mechanical assistance, such as an actuator or operator, to operate the valve.

Internal leak: A leak within the pipeline from one side of a valve to the other; this may be leak-by or leak-through.

Leak-by: For double-seated valves, an internal valve leak. It is a condition in a gate or ball valve in which hazardous liquid can leak past either the upstream or downstream seal into the valve body, thereby pressurizing the valve body.

Leak-through: A condition in a gate or ball valve in which hazardous liquid can leak past both valve seats, causing the valve to leak from the high-pressure side of the valve to the low-pressure side when the valve is closed. For single-seated valves, such as check valves, it is a condition where hazardous liquid can leak by the liquid seat, causing the liquid to internally leak past the valve when the valve is in the closed position.

Packing: The material between the valve stem and bonnet used to provide a leak proof seal around the stem, preventing fluid from leaking up around the stem.

Port: The internal opening for flow through a valve.

Seat: The nonmoving part of a valve upon which the disc rests to form a seal and close the valve.

Stem: The part of a valve that connects the valve operator to the disc. Stems can have rotary, helical, or linear movement.

Throttling: Reducing or controlling the fluid flow in a pipeline.

Travel: The movement of a stem as it positions a valve to either the open or closed position.

Trim: A collective term that includes the internal parts of a valve that receive the most wear and can be replaced. The trim includes the stem disc, seat ring, disc holder or guide, wedge, and bushings.

CONTREN™ LEARNING SERIES — USER UPDATES

The NCCER makes every effort to keep these textbooks up-to-date and free of technical errors. We appreciate your help in this process. If you have an idea for improving this textbook, or if you find an error, a typographical mistake, or an inaccuracy in NCCER's Contren™ textbooks, please write us, using this form or a photocopy. Be sure to include the exact module number, page number, a detailed description, and the correction, if applicable. Your input will be brought to the attention of the Technical Review Committee. Thank you for your assistance.

Instructors – If you found that additional materials were necessary in order to teach this module effectively, please let us know so that we may include them in the Equipment/Materials list in the Instructor's Guide.

Write: Curriculum Revision and Development Department
National Center for Construction Education and Research
P.O. Box 141104, Gainesville, FL 32614-1104

Fax: 352-334-0932

E-mail: curriculum@nccer.org

Craft ______ Module Name ______

Copyright Date ______ Module Number ______ Page Number(s) ______

Description ______

______

______

______

(Optional) Correction ______

______

______

(Optional) Your Name and Address ______

______

______

Module 62204-02

Pipeline Coating Inspection

(CT 5)

Introduction to the Pipeline Operator Qualification Rule, API, and NCCER

It could be said that petroleum products keep America running. Petroleum products like oil and natural gas are the principal sources of energy in the United States and are used in many manufacturing and chemical processes. Petroleum products are primarily transported by pipeline. There are more than a million miles of liquid and natural gas pipeline in this country, ranging in diameter from two inches to four feet. These pipelines bring crude oil and natural gas from production fields to processing plants and refineries—and ultimately to the consumer.

Pipeline Operator Qualification

Petroleum products are considered hazardous, so pipeline personnel must take extra precautions and follow specific procedures when working with them. To help ensure the safe handling and transportation of petroleum products, the United States Department of Transportation (DOT) mandates safe practices as federal regulations.

One such regulation, the Operator Qualification (OQ) rule, requires pipeline operators to develop a written qualification plan to evaluate their personnel and their contract personnel (see *49 CFR Part 192 Subpart N—Gas,* and *49 CFR Part 195 Subpart G—Liquid*). The evaluation hinges on the operator's ability to perform certain covered tasks and to be able to recognize and respond to abnormal operating conditions that may be encountered while performing these covered tasks.

Covered Tasks

A pipeline performance task is deemed *covered* if the activity passes the following four-part test:

1. Is it performed on a pipeline facility?
2. Is it an operations or maintenance task?
3. Is it performed as a requirement of *49 CFR Part 195 (Liquid)* or *49 CFR Part 192 (Gas)*?
4. Does it affect the operation or integrity of the pipeline facility?

If the answer to all four questions is *yes,* then the activity is considered *covered* and must figure into the evaluation process of pipeline personnel.

American Petroleum Institute

The OQ rule and its related covered tasks provided the catalyst to develop a unified pipeline training and assessment program—so that pipeline operators and owners have ready access to the assessment and training tools they need. This effort was the direct result of the vision and leadership provided by The American Petroleum Institute (API) and API members.

Efforts to unify the industry are nothing new to API. Indeed, the impetus for forming the Institute in 1919 was the need to standardize engineering specifications for drilling and production equipment. Today, the most pressing issues revolve around public perceptions and government policies toward the industry. Speaking with one voice on these issues has become as essential as having interchangeable parts in the field. API's members determine the petroleum industry's positions on public policy as well as the standards that govern its day-to-day operations worldwide.

Consortium on Operator Qualification

API assembled a focus group whose primary purpose was to identify covered tasks and provide compliance recommendations to pipeline operators.

The Consortium on Operator Qualification (COOQ) became the guiding force for the identification and development of covered tasks for the liquid pipeline industry. Their work product, API publication 1161, *Guidance Document for the Qualification of Liquid Pipeline Personnel,* formed the basis for the resulting Pipeliner Training and Assessment Program.

The covered task list created in API publication 1161 has now evolved to include gas covered tasks and is maintained by NCCER. The most current covered task list can be downloaded at www.nccer.org.

National Center for Construction Education and Research

With the guidance and support of API, its members, and the Consortium on Operator Qualification, the National Center for Construction Education and Research (NCCER) was able to design and implement the Pipeliner Training and Assessment Program.

NCCER is a not-for-profit 501(c)(3) education foundation established in 1995 by the world's largest and most progressive construction and maintenance contractors and national contractor associations. NCCER was founded to address the

severe workforce shortage facing the construction and maintenance industries and to develop a standardized training process and curricula. Today, NCCER is supported by hundreds of leading construction and maintenance companies, manufacturers, and national associations and developed more than 35 construction and maintenance titles.

NCCER also maintains a National Registry, which contains the records of trainees and instructors who have received NCCER training. All completion information for every trainee is sent to the NCCER and filed within the Registry. The Registry can then confirm training and skills for workers as they move from company to company, state to state, or even within their own company.

Standardized Curriculum (NCCER's Contren™ Learning Series)

NCCER curricula are developed using the expertise of industry subject matter experts (SMEs). The use of SMEs as an integral part of every new development or revision ensures that the curriculum is current, industry-driven, and indicative of industry practices. The development and/or revision process is facilitated from start to finish by NCCER project managers, who work closely with the SMEs and a select team of technical writer(s) on the development and presentation of the curriculum.

Pipeline: Five Training Paths

The pipeline curriculum focuses on essential skills and competencies within five skill areas:

- Corrosion Control
- Electrical and Instrumentation
- Maintenance
- Mechanical
- Operations (Field and Control Center)

SMEs from the pipeline industry have determined the skills and competencies needed based on API's covered tasks and the consensus of the group/industry. Specific company information is generally not incorporated into the curriculum unless the SMEs determine that the information is the industry standard.

Training Modules and Covered Tasks

The smallest unit of instruction in the curriculum is a *module*. Each curriculum is composed of modules (the equivalent of stand-alone chapters) that are logically sequenced into complete *levels* of instruction. Every module is task-oriented and requires the successful completion of a written (knowledge-based) examination and a performance (hands-on) test. Modules may be individually selected and/or combined for efficient task training. Modules in the pipeline curricula, to the extent possible, will align with API-developed covered tasks. While not all modules will relate to covered tasks, all covered tasks will relate to at least one module.

Benefits of Training to the Pipeline Marketplace

Applying NCCER's standardized training process allows pipeline owners and contractors in the United States to establish the benchmark for acceptable skill and competency levels within the pipeline industry. This process will assist training organizations in developing and/or improving standardized pipeline training programs that effectively match their resources and operating conditions while responding to regulatory demands and to the needs of the industry. The process will also recognize training organizations that meet the minimum standards of competence, warranting regulatory, public, and professional confidence. This in turn will foster national unity within the pipeline industry for the mutual benefit of individual technicians, owners, contractors, and ultimately pipeline-users and society.

Additional Resources

To review the Pipeliner Training and Assessment Program Implementation and Resource Guide and the latest version of pipeline covered tasks, visit NCCER's Web site at www.nccer.org, or contact:

National Center for Construction Education and Research
PO Box 141104
Gainesville, FL 32614-1104
(888) 622-3720
Web site: www.nccer.org

For more information on API, please contact:

American Petroleum Institute
1220 L Street, Northwest
Washington, D.C. 20005
(202) 682-8000
Web site: www.api.org

To order *API Pub 1161*, contact Global Engineering Documents by calling 1-800-854-7149 or visiting their Web site at www.global.his.com.

Additional sources of information on the pipeline operator qualification regulation include:

DOT – Office of Pipeline Safety
www.ops.dot.gov

Transportation Safety Institute
www.tsi.dot.gov

COURSE MAP

This course map shows all of the modules in the second level of the Pipeline Maintenance curriculum. The suggested training order begins at the bottom and proceeds up. Skill levels increase as you advance on the course map. The local Training Program Sponsor may adjust the training order.

PIPELINE MAINTENANCE LEVEL TWO

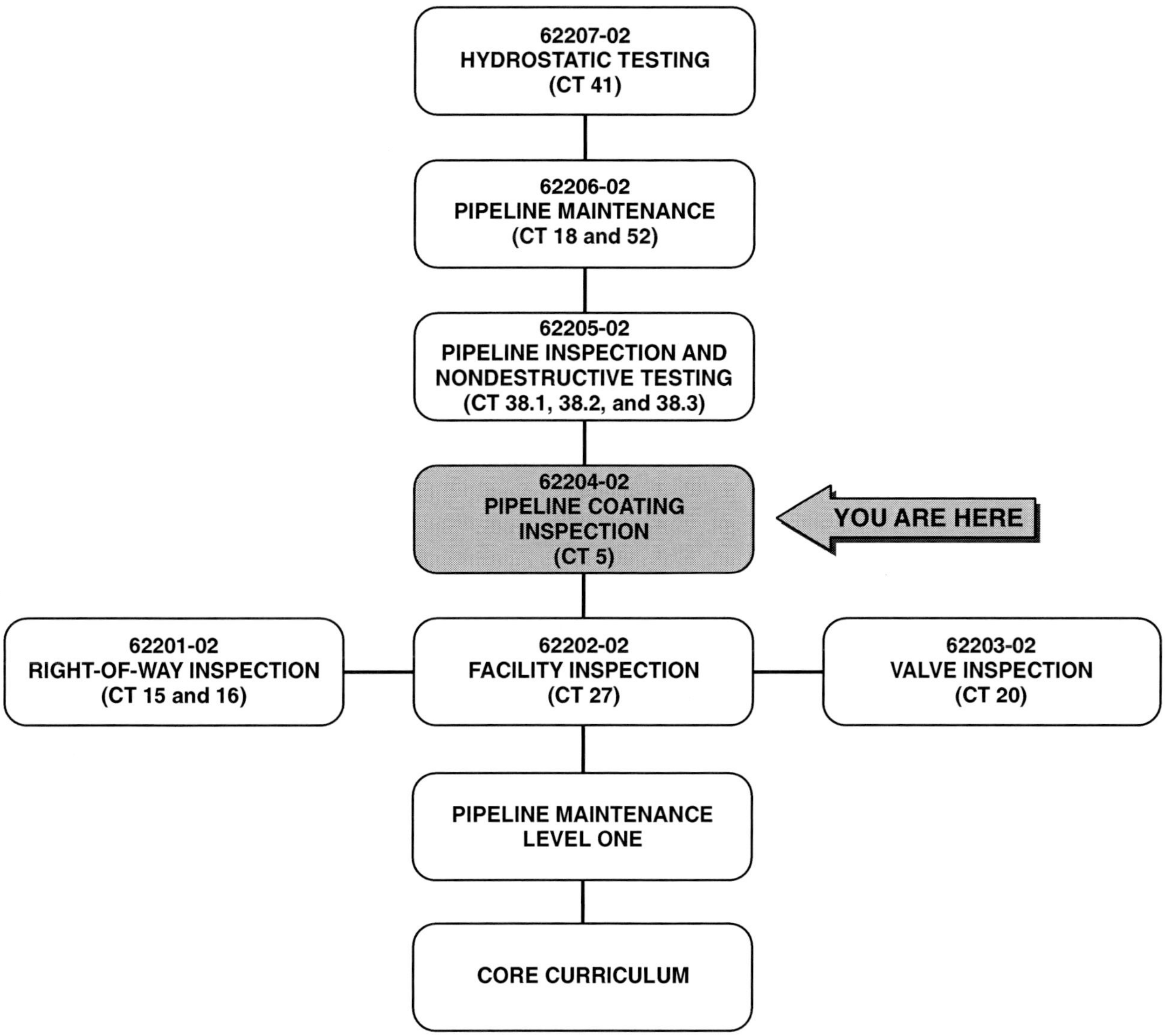

205CMAP.EPS

MODULE 62204-02 CONTENTS

Figures

Table

MODULE 62204-02

Pipeline Coating Inspection (CT 5)

Objectives

When you have completed this module, you will be able to do the following:

1. Identify pipeline coating types, and describe characteristics.
2. Identify pipeline coating problems (CT 5.1).
3. Inspect pipeline coating for damage (CT 5.3).
4. Inspect pipeline coating for external corrosion (CT 5.2).

Prerequisites

Before you begin this module, it is recommended that you successfully complete the following: Core Curriculum; Pipeline Core, Modules 66101-02 and 66102-02; Pipeline Maintenance Level One; Pipeline Maintenance Level Two, Modules 62201-02 through 62203-02.

Required Trainee Materials

1. Pencil and paper
2. Appropriate personal protective equipment

1.0.0 ◆ INTRODUCTION

Corrosion occurs as the result of a voltage differential between two components of a system. An anode and a cathode must be present. If a complex structure is constructed of materials that are near each other in the galvanic series, then there is reduced likelihood of potential differences. Construction using materials that are not close in the galvanic series, such as steel and copper, can create a corrosion cell. Nonmetallic materials also play an important role in corrosion control.

Coating and wrapping pipe is the main line of defense against pipeline corrosion. The primary purpose of coatings is to insulate the pipe from the environment. Any moisture penetrating the coating can carry electrical current to the pipe, initiating the corrosion process. For this reason, resistance to water penetration is a coating's most important characteristic. Unfortunately, coatings are not perfect, and accelerated corrosion occurs at breaks, or **holidays**, in a coating. Cathodic protection is used to prevent corrosion of the pipeline surface at the holidays in the coating.

Coating also plays an important role in cathodic protection engineering. Although it is possible to protect a bare structure, if the structure is coated, it is necessary to protect only the metal exposed at the holidays. This greatly reduces the size and cost of cathodic protection systems.

Coating inspection is important because it is desirable that the coating be in the best possible condition. This will ensure that the cathodic protection system performs as designed.

This module covers the following:

- Pipeline coating
- Pipeline coating problems
- Pipeline coating inspection

2.0.0 ◆ PIPELINE COATING

Pipe coating is a special material applied to the pipe exterior surface to prevent the environment from coming into contact with the steel. The initial pipe coating is applied to the pipe either at the mill or in the field during construction or maintenance. Pipe ends that are to be welded are not coated until after the welding is done.

There are a variety of coating types, and the coating selected for a particular pipeline in a particular location is chosen based on several factors. One factor is the environment and the soil type in which it is to be used. Cost, and company policy and the types of storage and handling the pipe will receive before being buried, are other factors.

Once the coating is selected, it must be applied correctly to be most effective. During pipe installation and maintenance, it must be inspected for flaws and protected from mechanical and rock damage. Different types of coatings have specific characteristics.

The following sections discuss types of coating, coating characteristics, and coating application.

2.1.0 Types of Coating

Different coating types are available to meet the specific demands of the pipeline. Some of the more commonly used coatings are:

- Coal-tar (bituminous enamels)
- Fusion-bonded epoxy
- **Tape coatings**
- Extruded polyolefin and polyethylene
- Two-component liquid resin coating
- Girth weld and other field coating

2.1.1 Coal-Tar (Bituminous Enamels)

Coal-tar is derived from pitches or petroleum asphalts and is reinforced with glass and a felt wrapper to provide additional mechanical strength and impact resistance. The exterior of the pipe is wrapped with kraft paper for protection during shipping and handling. Enamel-coated pipe must be handled carefully to prevent damage to the coating. These coatings have been in wide use since the early 1900s but are being used less often today because of toxicity and environmental concerns.

Enamel coatings include asphalts, coal-tars, greases and waxes, mastics, and asphalt mastics. Asphalts occur naturally and as a petroleum distillation by-product. Coal-tars are formed from coal distillations. Grease and wax are also petroleum based and contain corrosion inhibitors and fillers.

Mastics are mixed from asphalt, sand, and fillers and provide thicker, more durable coatings with greater adhesion. Asphalt mastics have been used for a long time and are mixtures of sand, crushed limestone, and fiber and contain an asphalt binder. This is the thickest coating type and is often used offshore.

All enamel coatings are unsuitable for use in hydrocarbon-contaminated soil. In very colder temperatures, enamels require special precautions to prevent cracking.

2.1.2 Fusion-Bonded Epoxy

Fusion-bonded epoxy consists of a powdered resin is electrostatically sprayed onto a pipe heated to about 400° to 500°F. The typical thickness is 12 to 15 mils. These coatings have good adhesion to steel and resist impact and abrasion damage well. Long-term moisture absorption can be a problem with these coatings, so cathodic protection current requirements may increase over time.

Powdered **epoxy resins** form a virtual skin over the steel surface when applied to heated pipe. The fusion-bonding process requires specialized equipment and exacting preparation, so these coatings are usually applied at the mill.

One big advantage of the epoxy resin coating is its relative thinness. Inspection of the pipe is easier since defects are not hidden under thick layers of coating. Epoxy resins are also strong enough to resist handling damage, are durable and flexible, and can be applied to interior pipe surfaces. They can also withstand high operating temperatures.

2.1.3 Tape Coating

Most tapes are cold applied. The system consists of a primer, a corrosion sealant inner layer, and a mechanical protective outer layer backing, which is usually polyethylene or polybutylene. There are several multilayer systems available. Pipes that are tape coated must be shipped and handled with care to prevent damage. They are prone to **disbonding** if the tape is not properly tensioned.

Polyethylene tape is often used for field-applied coating. Other tapes used for coating include polyvinyl, coal-tar base, and butyl-mastic. An advantage of tape coating is that it does not raise any environmental or health objections. *Figure 1* shows a tape coating application.

2.1.4 Extruded Polyolefin and Polyethylene

These coating systems consist of polyolefin or polyethylene extruded over a butyl or asphalt mastic adhesive. The coatings have high impact resistance and resist the development of holidays over time.

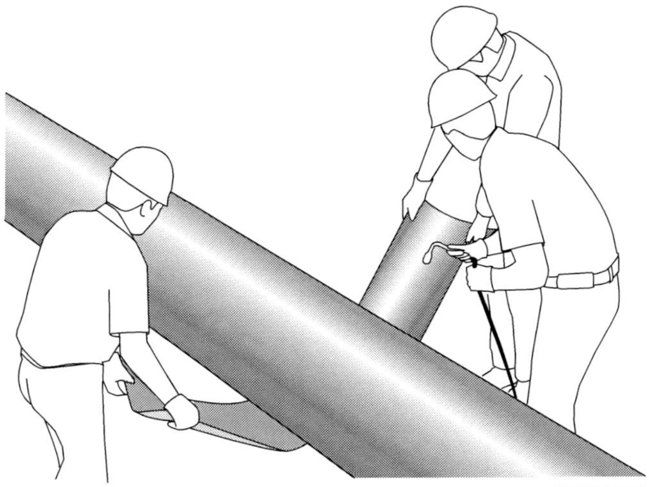

Figure 1 ◆ Tape coating application.

2.1.5 Two-Component Liquid Resin Coatings

These coatings consist of a liquid resin, usually an epoxy, and a curing agent. The two components are sprayed in the correct proportion onto the pipe, usually in two coats. These coatings are resistant to impact and abrasion damage.

2.2.0 Coating Characteristics

Coatings are films of various materials that protect the metal surface of the pipe on which they are applied. Coatings are selected based onto specific strengths and characteristics to meet the needs of their intended application. The desired characteristics, or coating properties, of any material to be used as a coating include:

- Proper adhesion
- Mechanical resistance
- Ease of application

2.2.1 Adhesion

Adhesion to the pipe surface is critical in order for the coating to protect the pipe. Adhesion is created by the interaction of forces between the coating material and the pipe. These forces are both chemical and physical. A coating that has strong adhesion properties will often outlast other coatings that have less adhesion but are stronger in other properties. Lack of adhesion causes many types of serious coating damage. *Figure 2* shows the effect of adhesion on moisture transfer.

Obtaining proper adhesion depends on selecting the right coating for the job and following appropriate application procedures. Different types of adhesion include chemical, polar, and mechanical. Chemical adhesion is created by a chemical reaction between the coating and the pipe surface. Polar adhesion is caused by the attraction of the resin molecule to the substrate. Mechanical adhesion is associated with the surface roughness or anchor pattern of the substrate. Anchor pattern is the roughness formed by peaks and valleys in the pipe surface that effectively increases the surface area for bonding.

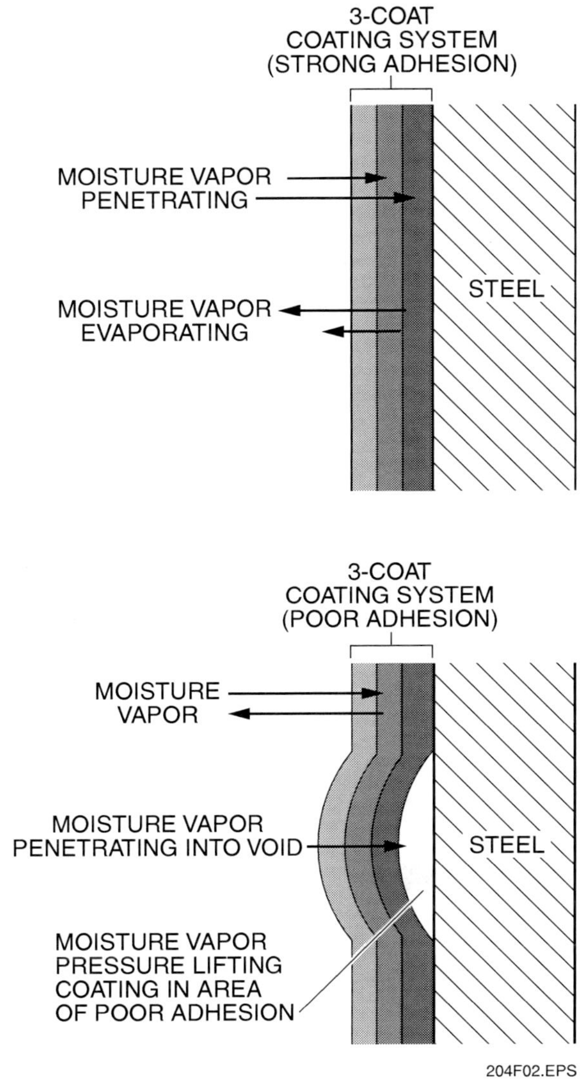

Figure 2 ◆ The effect of adhesion on moisture transfer.

2.2.2 Mechanical Resistance

Pipeline coatings must be resistant to a number of influences. Resistance to water is probably the most important property of an effective coating because water strongly affects almost all materials. Water can attack coatings in a number of ways, including absorption, moisture vapor transfer, thermal gradient across the coating, osmosis, and **electroendosmosis**, an effect of coated pipe being under cathodic protection. No single coating can be resistant to water in all circumstances.

Dielectric strength is the ability of the coating to break the electrical circuit set up during a corrosion reaction and is a key characteristic for coatings used with cathodic protection. Dielectric strength is affected by moisture absorption. The lower the moisture absorption of the coating, the higher the dielectric strength.

Resistance to chemicals to which the coating may be exposed in its environment is also critical. Chemicals that may be present include salts, acids, alkalis, and organic chemicals, such as petroleum products. Resistance to alkalis is extremely important in the cathode area of a protection system because development of alkalis will undercut the coating and spread corrosion.

Abrasion resistance may be an important property; this depends on the pipeline environment after installation. It is important in protecting the coating and pipe during the installation process. Abrasion can be caused by sand scouring, or by rocks, and by tools and equipment. Abrasion can result in undercutting of the coating. Undercutting is the action of rust forming under the coating, usually around a small break; it can also occur when the coating is applied over a corroded surface or contamination. *Figure 3* shows undercutting caused by abrasion.

Weather resistance is necessary for aboveground pipeline structures. Weather conditions that must be protected against include sunlight, hail, rain, snow, dew, freezing, and thawing. Temperature variations can cause contraction and expansion of the substrate under the coating. Extreme temperature variations can affect the adhesion and other properties of the coating. Coatings that can expand and contract with temperature may be able to withstand these extremes.

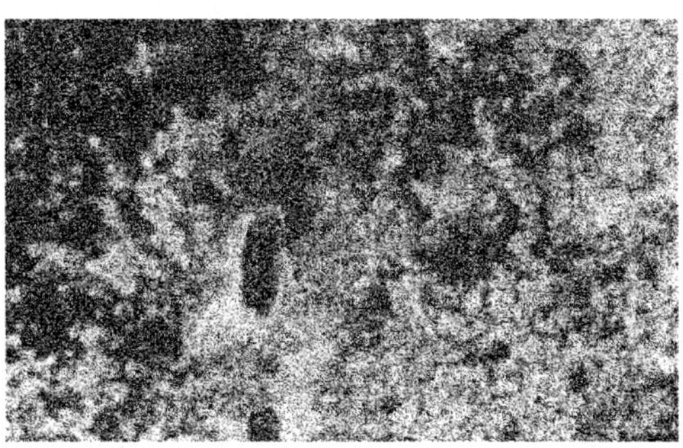

204F03.EPS

Figure 3 ◆ Undercutting caused by abrasion.

Atmospheric effects that may also be included under the category of weather are chemical fumes, dust, and particulate fallout.

Resistance to soil stress is particularly important for buried pipe. As soil contracts and expands due to moisture or temperature variations, the soil can pull and push at the coating surface, pulling it away from the pipe and causing cracks, voids, and thin spots. Coatings used in high-clay soils are most threatened by soil stress. Other soil stress factors may be caused by poor backfill methods. Rocks, roots, and uneven backfill pressure can damage the coating.

Resistance to dirt, bacteria, and fungi is important in many areas because these bacteria and fungi can settle in dirt deposits and actually attack the coating. These colonies can provide access for corrosion conditions. In some cases, coating materials and cements are used by the bacteria for food. Resins, pigments, and plasticizers are some materials that cannot be used by bacterial organisms for food.

The physical strength of a coating is a factor in its ability to resist abrasion, mechanical damage, and soil stress. The coating must have the strength and durability to maintain protection effectively over a period of many years under widely differing conditions.

2.2.3 Ease of Application

Pipeline structures and miles of pipelines offer challenging conditions for coating applications. It is important that the coating be as easy as possible to apply, but it must meet criteria and requirements of the specific application. This will help ensure that the application is done correctly. If application is difficult to accomplish, welds, bends, fittings, or even the lengths of pipe may suffer **coating flaws** that will invite corrosion attack.

2.3.0 Coating Application

The effectiveness of a coating is dependent not only on its characteristics but also on using proper application procedures and protecting the coating during and after installation. There are many opportunities during installation of the coated pipe for damaging the coating. The environment in which it has been placed may also have a detrimental effect on the coating. If the coatings are compatible, almost any type of coating can be used on field joints or repaired sections. Proper coating application includes preparation, application, and protection.

2.3.1 Preparation

Ensuring that the pipe is properly prepared for coating application is extremely important. The pipe surface must be free of all moisture, dirt, scale, rust, and debris. If it is not, the coating will not adhere properly and will not be effective. Each type of field coating has specific surface preparation requirements that must be followed. For all coatings, the surface must be clean and dry. Often this is accomplished with power tools, but in some cases, abrasive blasting and solvents are required. In the mill or the field, brushes and buffers may be used.

Some coatings require a primer on the prepared surface prior to the application of the coating. Primer application must be done in accordance with the manufacturer's specifications and must be allowed to dry to the specified consistency before the outer coating is applied. The coating and wrapping are done in accordance with manufacturer's specifications. In the field, special machines are used for cleaning, priming, coating, and wrapping directly over the ditch in which the pipe is being laid.

2.3.2 Application

Specific coatings have specific application requirements and procedures. Coatings may be applied hot or cold depending on the type of coating. The common field coatings in use today are liquid epoxies, heat shrink sleeves, and various types of tape. All coatings should be applied in accordance with the manufacturer's requirements.

Enamels are melted in **dope pots** and kept at the right temperature for use in coating. The liquid coating is stirred constantly to evenly distribute heat and to keep the liquid uniform before it is fed through a hose to the coating machine, where it is applied in an even layer. Wrapping is applied over the coating to protect it while it hardens.

Fusion-bond epoxy coatings require the pipe to be heated and also require the use of special equipment and exacting procedures. Because of this, fusion-bond epoxy coatings are not normally applied in the field.

Taping machines apply tape in the field by moving along the line and wrapping tape around the pipe. Tape is supplied from rolls mounted on spindles on the machine, fed to the pipe under tension, and wrapped in overlapping layers.

Because the ends of pipe to be welded are left uncoated, they must be coated in the field after welding. The same applies to parts of the pipe from which coating has been removed for repairs. A variety of processes and materials may be used. If the rest of the pipe is coated with coal-tar, then coal-tar, glass, and felt are usually applied to the weld joints or repaired section. Heat-applied epoxy powder can be used with thin film coatings. It is critical that the joint with adjacent coating be a waterproof joint.

2.3.3 Protection

Pipe wrapping is the material applied on top of the coating to protect it from damage during handling. Wrapping materials are usually polyethylene, polyolefin, felt, fiberglass, or kraft paper. Kraft paper also protects pipe from ultraviolet rays if it is exposed to the sunlight but is used only with coal-tar coatings.

Other factors that pose a hazard to the coating include the moisture content of the soil, contamination of the soil, soils that vary greatly from wet to dry, and extreme temperature variations. These factors must be addressed when selecting the type of coating to use and the type of wrapping to cover it.

In some areas, it is difficult to remove all rocks from the environment. In these cases, **rock shields** can be used for protection. Rock shields must be cathodically protected. Some type of protective barrier may also be used to protect the coating from damage when existing pipe is excavated.

3.0.0 ◆ PIPELINE COATING PROBLEMS

Coating problems may be flaws that occur if the coating is not applied correctly, improper coating is used, or the pipe is not prepared properly. Flaws can also be the result of not using the correct coating for the job. Coating damage may occur during application or installation or after the pipeline is put into service. Coating that has become damaged in place or during pipe repairs must be removed and replaced. The replacement coating must meet or exceed the original coating requirements. *Figure 4* shows an example of coating failure.

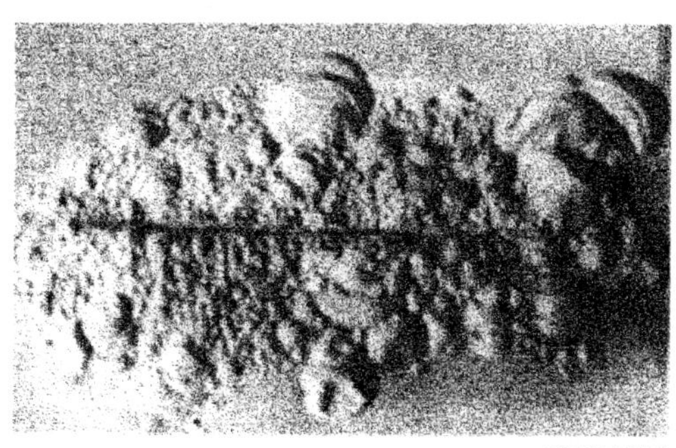

204F04.EPS

Figure 4 ◆ Example of coating failure.

3.1.0 Common Coating Failures

Coating failures can vary widely depending on conditions. Some common pipeline coating failures may include:

- Disbonding
- Incorrect thickness
- Wrinkles
- Foreign inclusions
- Incompatibility

3.1.1 Disbonding

A common type of coating failure is the disbonding, or separation, of the coating from the pipe (see *Figure 5*). Disbonding is lack of adhesion. The cause for lack of adhesion depends on the type of adhesion. The problem can usually be traced to improper application or selection of the coating or poor handling and preparation procedures. In some cases, however, the pressure differential created by a voltage imbalance between the pipe and the soil electrolyte can build to the point that water is actually forced to migrate through the coating and onto the surface of the pipe. Disbonding of the coating and the beginning of a corrosion cell is the usual result of this process.

204F05.EPS

Figure 5 ◆ Disbonding.

3.1.2 Incorrect Thickness

When liquid coating is applied too thinly, there may not be enough material left on the surface to effectively coat it. Coating that is too thin can give rise to pinpoint corrosion. The pinpoints will gradually become larger until the entire area is undercut in the thinnest spots.

When liquid coating is applied too thickly, runs and sags in the coating can result. Thick coating may not dry properly; the surface dries much more quickly than the underlying body of the coating, and there may be an excessive retention of solvent. Retained solvents can cause blistering and poor adhesion. Internal shrinkage caused by excessive thickness can result in checks, cracks, and scaling, which leads to undercutting.

3.1.3 Wrinkles

If a liquid coating is applied too thickly, the surface and body of the coating will dry at different rates and cause wrinkles (see *Figure 6*). Wrinkles can also occur when excess surface driers are used. This problem occurs most frequently with oil-based coatings. Temperature can be a major factor in wrinkling. A higher curing temperature causes the surface to dry more quickly, and cold weather application can cause a film to develop on the surface.

Tape wrapping can also have wrinkles from incorrect application or from soil stress after the pipe is installed.

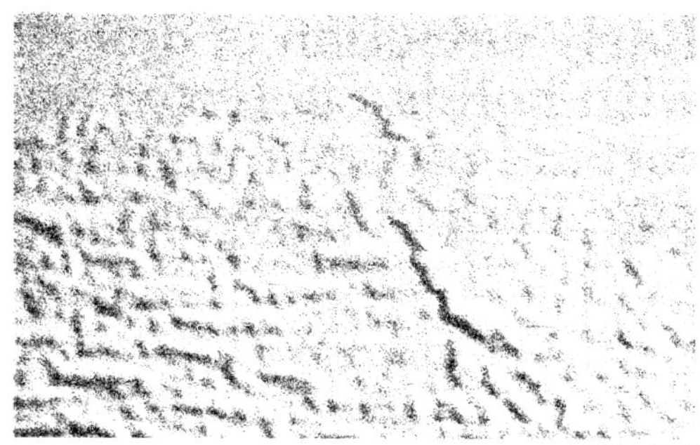

204F06.EPS

Figure 6 ◆ Wrinkling.

3.1.4 Foreign Inclusions

Any foreign inclusions in the coating material form weak points in the coating integrity. Depending on the nature of the inclusion and the type of coating, the foreign matter can also degrade the coating itself. This will prevent proper adhesion and provide access to corrosive conditions.

3.1.5 Incompatibility

The coating selected for the application must be compatible with any existing coating. Mixing of different types of coating on one application can cause lack of adhesion at joints or negative effects of one material on the other. All components of the coating system must be compatible. For example, the tape wrapping must be compatible with the primer.

3.2.0 Pipeline Coating Damage

Damage to the pipeline coating can occur as a result of handling or excavation procedures, environmental factors, or contact with other materials. Causes of pipeline coating damage may include:

- *Soil stress* – Moisture in the soil around the pipe can lead to soil stress in the coating. Soils that are alternately wet and dry can be abrasive against the coating surface and can stress it to the point of failure.
- *Foreign chemicals* – Foreign chemicals in the soil, such as hydrocarbons, can cause degradation of some coatings, particularly enamels
- *UV damage* – When pipe is aboveground, ultraviolet light can be damaging to some coating surfaces.
- *Rock damage* – Rocks in the soil can damage the coating and the pipe. Rocks in the backfill can scrape, dent, or gouge the pipe coating when the ditch is filled. Rocks under the pipe can cause pipe bending or ruptures when the weight of the pipe and liquid presses down. Care is usually taken to remove rocks from the ditch and backfill material to avoid damage to the pipe and coating. Rock shields are also used to protect the pipe and coating from rock damage.
- *Mechanical damage* – Mechanical damage can be caused during excavation handling, installation, storage, or transportation. This damage includes dents, scrapes, cuts, cracks, gouges, and tears.

4.0.0 ◆ COATING INSPECTION

Coating must be inspected when it is applied and whenever the pipe is exposed during service to ensure that it is performing as it should. Coating is inspected visually and by machine.

4.1.0 Visual Inspection

During the coating application, surface preparation must be inspected to ensure adherence with the specifications. Environmental conditions, such as temperature, humidity, and dew point, must be checked against the manufacturer's or company's specifications. This is to ensure the surface is being prepared and the coating is being applied in an atmosphere conducive to proper coating performance.

While detailed inspections need to be made by the coating inspector, cathodic protection personnel and maintenance personnel may be called on to inspect exposed coating. This can be done visually by looking for signs of disbonding, damage, checking, cracking, chalking, and other deterioration. The purpose of this inspection is to determine the condition of the coating to see if any recoating is required. If a poor coating is found, it may be necessary to excavate more of the pipe to see how far the deterioration exists.

4.2.0 Machine Inspection

The machine used to inspect coatings is called a **holiday detector**. A holiday is a gap or void in the pipeline coating or in the paint on a metal surface. A holiday detector is an electrical device used to locate a thin spot or void in coatings on pipelines and fittings. It is also called a jeep. These instruments are made in high-voltage and low-voltage versions. The low-voltage detectors are used only for thin film coatings that are less than 20 mils thick.

The machine consists of a metallic ring that encircles the pipe, a handheld voltage source, and an alarm unit. The detector applies a voltage between the coating and the pipe. When a holiday is encountered, the probe shorts to the structure, a spark occurs, and an audio alarm sounds. The alarm or buzzer indicates the location of a holiday, which is marked for repair.

> **WARNING!**
> Because of the spark, holiday detectors must not be used in an explosive or flammable atmosphere. Because high voltage is often involved, precautions must be taken to avoid electrical shock.

When defects are found, they are marked on the surface and repaired. After repairs are made, the coating is inspected again.

Another type of detector, the Pearson holiday detector, is passed over the line after the pipe is buried to detect coating defects and the presence of any metal debris. Metal debris near the line may cause future corrosion problems. Coating flaws discovered during inspection can be manually repaired depending on the extent of the problem. The Pearson survey is a fast way to locate holidays. Since holiday detection is continuous over the entire length of the structure, there is no danger of missing one. To perform a Pearson survey, the two operators are fitted with shoe cleats that are connected with 30 feet of cable. The transmitter is connected to the structure, and one operator precedes the other walking directly over the line. The person in the rear handles the cable to keep it off of the ground and clear of obstructions. The person in front carries the amplifier connected to the earphones.

The National Association of Corrosion Engineers (NACE) International has recommended practices that cover holiday testing. Documents cover high- and low-voltage tests and list proper voltages for different coatings. It is important that the correct detector and the correct voltage be used for the coating involved. If too low a voltage is used, the holiday may not be found. Too high a voltage may damage the coating.

A good way to determine if the voltage is sufficient and to see if the pipe is properly grounded is to create a tiny holiday in the coating using an ice pick or the point of a knife blade. Always follow manufacturer's procedures to select proper voltage. If the instrument is properly adjusted, a spark will occur at the known holiday but will not occur where the coating is of proper thickness, and there will be no burn-through of the coating.

4.3.0 Inspection Procedures

Visual coating inspections may be performed on aboveground pipe or on exposed buried pipe. Inspections must be documented on the proper company forms. *Figure 7* shows a sample inspection form.

4.3.1 Aboveground Pipeline Coating Inspection

Visual inspection of aboveground pipeline coating requires that coating be checked for damage caused by exposure. Follow these steps to inspect aboveground pipeline coating:

Step 1 Test the atmosphere for any hazardous gases to ensure that the area is safe for work.

Step 2 Ensure that the area is properly marked and required safety equipment is in place.

Step 3 Visually inspect the pipe for any signs of wetness or leaks.

Step 4 Inspect the soil below the pipe for any discoloration or other signs of contamination.

Step 5 Visually inspect the pipe for signs of pitting on the surface of the pipe.

Step 6 If present, determine the extent of any pitting and document the results. *Figure 8* shows the American Society for Testing Materials (ASTM) ratings of pits on a steel surface. *Table 1* shows the scale and description of ASTM rust grades.

NOTE

The use of the images in *Figure 8* and the use of *Table 1* can help the inspector to determine the extent of pitting or other damage in order to report it accurately. These images and *Table 1* are from *ASTM Standard D610-01*. The numbers over the images correspond to those in *Table 1*, except that 0 and 10, as extremes, are not illustrated.

Step 7 Visually inspect the pipe for signs of metal loss.

Step 8 If loss has occurred, use the appropriate instrument to measure the amount of loss, and document the results.

Step 9 Inspect the pipeline for general corrosion problems, including connections and welds.

Step 10 Document any corrosion findings.

Step 11 Visually inspect the pipe for coating flaws or damage, and document any findings.

Step 12 Identify any areas of thin spots or holidays in the coating, and document any findings.

Step 13 If the coating is to be repaired or replaced, retrieve a sample of the coating and any by-products under it for analysis.

Step 14 Complete the appropriate inspection form per company procedures.

4.3.2 Exposed Buried Pipeline Coating Inspection

Buried pipeline coating is only inspected when it is exposed. Pipeline is not excavated for the purpose of inspection. Follow these steps to inspect exposed buried pipeline coating:

Step 1 Ensure that the areas of the pipeline system have been properly excavated without damage to the pipeline.

Step 2 Ensure that the atmosphere above and in the trench has been tested for hazardous gases.

WARNING!

In case of the ignition of a hydrocarbon resulting in a fire or explosion, evacuate the area, and notify the fire department.

PIPELINE INSPECTION REPORT

Frequency: At Time of Inspection

PART A — LOCATION

Reason For Inspection ______________________

1. Location: Division ____________ County ____________ State ______ Date of Inspection ____________

A) **Mainline**
Alignment Sheet ________ Pipeline Size/Segment ________
Beg. M.P. ________ End M.P. ________
Beg. Sta. No. ________ End Sta. No. ________
Note: Station numbers shall be measured from a known landmark (Road Crossing, etc.)

B) **Company Facility: Pump Station, Tank Farm, etc.**
Location Name ____________
Physical Location of Pipe Inspected ____________
Milepost ____________

PART B — MATERIAL/CONDITIONS

2. Piping:
Pipe Size ________ In. O.D. Nom. W.T. ________ In. Actual W.T. ________ Grade ________

3. External Pipe Coating:

A) Coating
☐ Bare
☐ LAC 450
☐ Taped
☐ Insulated
☐ Other ________

B) Application
☐ Mill Coated
☐ Yard Coated
☐ Field Coated
☐ Other

C) Coating Thickness
Factory Applied____ mile
Field Applied ____mile

D) Condition of Coating
☐ Good ☐ Fair
☐ Poor

E) Bond
☐ Good
☐ Fair
☐ Poor

F) Cause of Coating Failure
☐ Damage
☐ Disbondment
☐ Defective Application
☐ Decomposition
☐ Other ________

4. Environment
☐ Clay ☐ Loam ☐ Sand ☐ Gumbo ☐ Caliche ☐ Rock ☐ Other ________
☐ Wet ☐ Dry ☐ Moist Depth of Cover (Inches) ________
Soil Resistivity Measurement ________ (ohm-cm) Measurement Method ________

5. Coating Repaired ☐ No ☐ Yes
Repair System ____________

6. Pipe-to-Soil Potential at Inspected Location (At Grade) ________ (volts) (At Pipe/Soil Interface) ________ (volts)

7. Were any Non-Destructive Tests Completed? ☐ No ☐ Yes Type ________

8. Evidence of: ☐ Corrosion ☐ Defect ☐ Damage ☐ Stress Corrosion Cracking (S.C.C.) ☐ Calcareous Deposits

9. Test Station Added ☐ No ☐ Yes I.D. No. ________

NOTES:
1) If S.C.C. or Corrosion is Noted, Complete Below and Back of Report (External Only, See Note 4)
2) If Defect or Damage is Noted, Complete Back of Report
3) If Repairs are Required, AAPL Form 502 "REPAIR REPORT FOR LEAK DAMAGE, CORROSION OR TEST FAILURE" Must be Attached.
4) AAPL Form 508 "INTERNAL CORROSION INSPECTION REPORT" Should be Completed When Internal Corrosion Is Detected.

SAMPLE

PART C — EXTERNAL CORROSION

10. General Corrosion Information:

A. DESCRIPTION
☐ Less than 10% of Nom. Wall Thickness, Complete Front of Report
☐ Greater than 10% of Nom. W.T., Complete Front and Back of Report
☐ Pitting ☐ General ☐ Localized
Max Depth ________ (In.) Longitudinal Length ________ (In.)

B. STATUS
☐ Active
☐ Inactive
☐ Unknown

C. CAUSE
☐ Galvanic
☐ Stray Current
☐ Unknown
☐ Other ________

11. Cathodic Protection: ☐ Impressed ☐ Galvanic

12. Pipe-to-Soil Potential:
a) Last Pipe-to-Soil Measurement at Nearest Point Each Side of Corroded Area
Upstream ________ (volts) Downstream ________ (volts)
b) Distance from Corroded Area to Measurement Points:
Upstream ________ (feet) Downstream ________ (feet)
c) Date of Measurements: Upstream ________ Downstream ________

13. pH of Soil Near Corroded Area ________ pH of Moisture Under Coating ________

14. Pipe Surface Temperature if Fluid is Heated ________ (deg F)

Remarks ____________

Distribution: (Original) Corr. Ctrl./Eng (Copy) Division Office (Copy) Other ________

________ *Technician or Inspector* __/__/__ *Date* ________ *Supervisor* __/__/__ *Date*

204F07A.EPS

Figure 7 ◆ Sample inspection form. (1 of 2)

PIPELINE INSPECTION REPORT

Frequency: At Time of Inspection

PART D — DEFECT, DAMAGE OR CORROSION INVESTIGATION

Notes:
1) Tabulate all Dents, Gouges, Slivers, Corrosion, S.C.C., etc.
2) List Reference Point - Valve, Tee, Center Road etc., as shown on Alignment Sheet with appropriate Station Number, if available.
3) Indicate Location of Longitudinal Seam. If Pipe is Vertical Substitute Compass direction for o'clock Position. For Spiral Weld Pipe, Note as Such.
4) Defects Such as Dents, Slivers, Gouges, etc., shall be tabulated individually. Corrosion Cracks and/or Corrosion Pits Closely Grouped May be Tabulated as One Defect.
5) Defect Width and Length Will be Measured Along the Girth and Longitudinal Axis Respectively.

Reference Point: ☐ Upstream End (Station No.) ____________ Description ____________

☐ Downstream End (Station No) ____________ Description ____________

Upstream End **Downstream End**

SKETCH DEPICTING PIPE CUT AT BOTTOM AND LAID FLAT
NOTE: Indicate Girth Weld Location If Known

O'Clock Position →

← Length of Inspected Area →

PART E — DEFECT, DAMAGE OR CORROSION MEASUREMENTS AND LOCATION

No.	Type	Distance from Girth Weld	Distance from Long Seam	Width (Circumferencial)	Length (Longitudinal)	Maximum Depth	Number of Anomalies
1.							
2.							
3.							
4.							
5.							
6.							
7.							
8.							
9.							
10.							
11.							
12.							
13.							
14.							

If Repair is Required, send to Division Manager for Approval.

Repair Required ☐ No ☐ Yes Type of Repair ____________

Approved by ____________ Title ____________

Remarks: ____________

204F07B.EPS

Figure 7 ◆ Sample inspection form. (2 of 2)

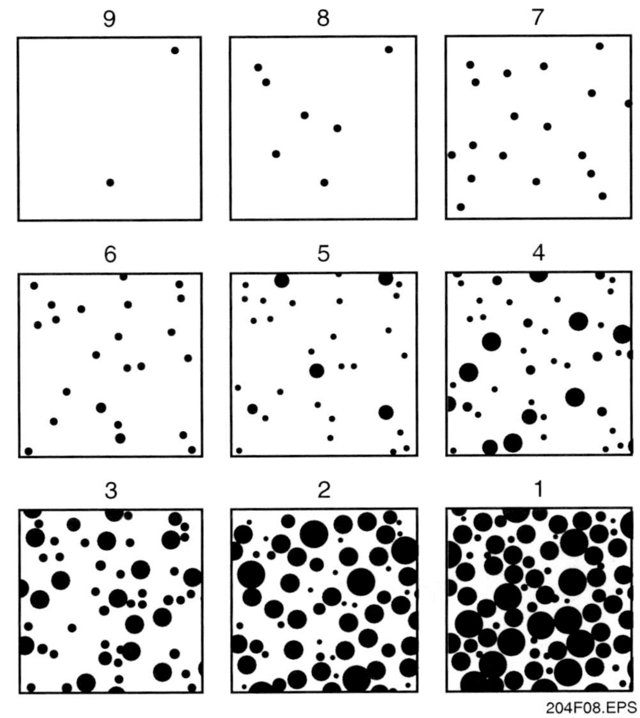

Figure 8 ◆ ASTM ratings of pits on a steel surface.

Step 3 Ensure that the trench has been properly constructed and that safety procedures are in place.

Step 4 Visually inspect the pipe for any signs of soil discoloration, wetness, or fungus growth before cleaning the pipe.

Step 5 Take samples of any signs of contamination, and document findings.

Step 6 Visually inspect the pipe for any signs of wetness or leaks.

WARNING!

If you encounter unexpected carbon dioxide or a hazardous liquid coming from a leak in the exposed pipeline, eliminate all ignition sources and notify the appropriate personnel.

Step 7 Inspect the soil below the pipe for any discoloration or other signs of contamination.

Step 8 Inspect the pipeline, including connections and welds, for general external corrosion problems.

Step 9 Locate and identify any areas and signs of corrosion on the pipe, including pitting. Document any findings.

Table 1 Scale and Description of ASTM Rust Grades

ASTM Rust Grades	ASTM Description	Alternative Description
10	No rusting or less than 0.01% of surface rusted	No corrosion
9	Minute rusting, less than 0.03% of surface rusted	1 to 2 pinpoints of rust per square foot
8	Few isolated rust spots, less than 0.1% of surface rusted	3 to 6 pinpoints of rust per square foot
7	Less than 0.3% of surface rusted	± 20 pinpoints of rust per square foot
6	Extensive rust spots less than 1% of surface rusted	± 30 pinpoints to 1/32 in. rust spots per square foot
5	Rusting to the extent of 3% of surface rusted	± 30 pinpoints to 1/16 in. rust spots per square foot
4	Rusting to the extent of 10% of surface rusted	± 50 1/32 or 1/8 in. rust spots per square foot
3	Approximately 1/6 of surface rusted	Approximately 1/4 of surface rusted with 1/32 or 1/8 in. rust spots
2	Approximately 1/3 of surface rusted	Approximately 1/3 of surface rusted with 1/32 or 1/4 in. rust spots
1	Approximately 1/2 of surface rusted	Approximately 1/2 of surface rusted with 1/32 or 1/2 in. rust spots
0	Approximately 100% of surface rusted	Approximately 100% of surface rusted

204T01.EPS

Step 10 When corrosion is evident, measure the longitudinal axis of the corroded area(s) to determine the length of the corrosion.

Step 11 Note the location of the corroded area on the pipe, such as the top, bottom, or sides, and record all the information per company policy.

Step 12 Visually inspect the pipe for signs of pitting on the surface of the pipe. Determine the extent of any pitting. Using a pit depth gauge, measure the depth of the pits, and document the results. *Figure 8* shows ASTM ratings of pits on a steel surface.

> **NOTE**
>
> The use of the images in *Figure 8* and the use of *Table 1* can help the inspector to determine the extent of pitting or other damage in order to report it accurately. These images and *Table 1* are from *ASTM Standard D610-01.* The numbers over the images correspond to those in *Table 1*, except that 0 and 10, as extremes, are not illustrated.

Step 13 Visually inspect the pipe for signs of metal loss.

Step 14 If loss has occurred, use the appropriate instrument to measure the amount of loss, and document the results.

Step 15 Visually inspect the pipe for mechanical damage, such as dents, scrapes, scratches, and gouges. If found, determine the cause of the mechanical damage, document the extent of the damage, and notify those responsible for correcting damage. Mechanical damage is generally caused by equipment during general construction work, equipment during farming activities, and improper pipeline construction.

Step 16 Determine the shape and depth of the physical/mechanical damage in order to determine the method of repair.

Step 17 Determine the proximity of the dent in relationship to the seam and girth weld, measure the depth of the dent, and document any findings.

Step 18 Identify areas of the exposed pipe that have coating.

Step 19 Identify the type of existing coating. It is necessary to identify the correct type of coating on the pipe so that the proper selection can be made for repair to ensure compatibility, adhesion, mechanical integrity, and proper temperature range.

Step 20 Visually inspect the pipe for coating flaws, abnormalities, or damage, and document any findings. Recognizing any physical damage to the coating, such as peeling, cracking, bubbling, and disbondment, is necessary to make an accurate assessment of the coating damage.

Step 21 Identify any areas of thin spots or holidays in the coating, and document any findings.

Step 22 Remove all loose and disbonded coating. The removal of such coating enables the proper inspection and evaluation of the underlying pipe. The removal of the defective coating and subsequent surface preparation of the pipe surface ensures good adhesion of the repair coating to the pipe.

Step 23 If the coating is to be repaired or replaced, retrieve a sample of the coating and any by-products under it for analysis.

Step 24 Identify possible causes of the coating failure. Considering the geographic location of the failed coating, inspect the environment for clues and causes. Look for signs of poor surface preparation/application, debris in the backfill, third-party encroachments, and the presence of foreign materials.

Step 25 Document all findings and make company notifications as applicable.

5.0.0 ◆ TERMS AND DEFINITIONS

The following sections highlight terms that are commonly used in pipeline maintenance work. It is important to be familiar with these terms and their definitions.

5.1.0 Types of Corrosion

Many different types of corrosion may be found on the pipeline. Some types of corrosion that you may encounter include:

- *Bacteriological corrosion* – Corrosion that is caused by the action of bacteria on metals (typically SRB and APB bacteria).
- *Galvanic corrosion* – Corrosion that is caused by electrochemical action that attacks metal installations, such as tanks, buildings, and pipelines. In pipelines, galvanic corrosion causes pitting.
- *Stray current corrosion* – Corrosion of metals in contact with electrolytes that is caused by a stray current. Stray current is a portion of a current that flows over a path other than the one intended.

5.2.0 Types of Coating

There is a large selection of coatings from which to choose to protect the pipeline. The environment in which the pipeline will be placed plays a large role in this selection. Types of coating include:

- *Asphalt coating (somastic)* – A mastic composition that, after application, is converted into a solid protective adherent film.
- *Coal-tar coating* – A paint-like coating for petroleum product pipelines that provides protection from both water and cathodic corrosion. It is also called coal-tar enamel. Coal-tar is made from a tar obtained from the carbonization of coal, usually accomplished in coke ovens or retorts. Coal-tar contains several hundred organic chemicals. A coal-tar coating may have a felt or mesh outer wrap.
- *Extruded coating (extrucoat)* – A factory-applied coating that is made of an extruded plastic, such as polyethylene or polyolefin.
- *Liquid epoxy coating* – A material consisting of two parts that, when combined, provide a high-strength, low-shrinkage coating.
- *Petroleum coating product* – Rust-preventing products that contain petrolatum, which is a smooth, semisolid blend of mineral oil with waxes crystallized from residual-type petroleum lubricating oil. The wax molecules contain 30 to 70 carbon atoms and are straight chains with a few branches or napthene rings.
- *Primer* – A coating material intended to be applied as the first coat on an uncoated surface. The coating is specifically formulated to adhere to and protect the surface of the pipe and to produce a suitable surface for subsequent coats.
- *Tape wrap* – A wrap made of polyethylene or polyvinylchloride that is applied to the piping over a primer coat.

5.3.0 Applying the Coating

Many factors and techniques come into consideration when applying the coating. These include:

- *Bonding/adhesion* – The attachment of a coating to a substrate. Two types of adhesive bonds are chemical and mechanical.
- *Cigarette wrapping* – An application technique for cut lengths of sheet coating. The width of the sheet must exceed the circumference of the pipe to provide overlap. With the exception of heat shrink sleeves, this method is not commonly used.
- *Coating overlap* – The overlapping of each layer or application of a tape or cigarette coating to ensure the coating is well bonded to the next layer in order to prevent separation.
- *Hand tool surface preparation* – Preparation of the surface for coating using hand tools such as scrapers and wire brushes.
- *Preheating* – Heating the surface of a material and/or the material to a desired temperature before proceeding to the next step in the process.
- *Shrinking* – Contraction of a coating after being heated and allowed to cool.
- *Spiral wrapping* – An application technique for tape wrap coating. Tape is placed at a 20- to 45-degree angle along the girth of the pipe and overlapped according to the manufacturer's specifications.

5.4.0 Pipeline Surface and Coating Problems

When inspecting the coating of aboveground or exposed buried pipeline, you may encounter damage to the coating and surface of the pipeline. Damage that may be found includes:

- *Coating disbondment* – Failure of the bond between the coating and the pipe.
- *Dent* – A depression that produces a reversal in the curvature of the pipe wall without reducing the wall thickness.
- *Gouge* – Surface damage that reduces the pipe wall thickness by displacing or removing of metal.
- *Holiday* – An undesirable discontinuity or break in the anticorrosion protection (coating) of the pipe. Electronic testing devices detect flaws in the protective coating of the pipeline and joints as the pipe is being laid. When detected, the breaks are manually coated.
- *Pitting* – The metal loss on the external or internal surfaces of a pipe that appears in small, crater-like depressions; over time, becomes holes in the pipe.
- *Scratch* – A thin, shallow cut or mark on the surface of the pipe.

5.5.0 Miscellaneous

Additional terms with which you must be familiar include:

- *Pipe specification* – The size, shape, thickness, and coating type of a pipe.

- *Pit depth gauge* – Measures the depth of the depressions caused by corrosion and the remaining wall thickness.
- *Shrink sleeve products* – A shrink fit is an extremely tight fit resulting from shrinking the coating component around the structure. Heat is applied around a companion piece, and as it is heated, the coating component shrinks around the structure, resulting in a shrink fit.
- *Specified minimum yield strength (SMYS)* – Internal design pressure based on the diameter, thickness, and yield strength of the pipe.

Summary

Pipe coating inspection is a critical activity for maintaining a safe pipeline. Inspecting pipe coating involves understanding the coatings and identifying coating problems and damage. This module identifies the different types of coatings available and the characteristics of each type. It also explains how to identify the causes of coating failures and describes how to perform coating inspections. The types of coating inspection methods covered are visual inspections and inspections using equipment.

Review Questions

1. Corrosion occurs as a result of _____ differences between two components of a pipeline system.

a. pressure
b. temperature
c. moisture
d. voltage

2. Breaks in pipeline coating are called _____.

a. holes
b. holidays
c. blisters
d. scrapers

3. Pipe ends that are to be welded during installation are coated _____.

a. at the mill
b. before welding
c. after welding
d. after shipping

4. Fusion-bonded epoxy is heated to about _____°F.

a. 400 to 500
b. 500 to 600
c. 600 to 700
d. 700 to 800

5. Most tapes are applied _____.

a. hot
b. cold
c. wet
d. with rivets

6. The different types of adhesion include chemical, polar, and _____.

a. manual
b. wet
c. hot
d. mechanical

7. The most important property of an effective coating is resistance to _____.

a. minerals
b. electricity
c. water
d. air

8. Physical strength of a coating is important to resist all of the following *except* _____.

a. abrasion
b. water
c. soil stress
d. mechanical damage

9. Enamels are melted in _____.

a. kettles
b. ovens
c. dope pots
d. crucibles

10. The material applied on top of a coating to protect it is _____.

a. pipe wrapping
b. coal tar
c. primer
d. epoxy

11. Kraft paper is used to protect the coating from _____.

a. water
b. electricity
c. soil
d. ultraviolet light

12. Disbonding is caused by _____.

a. ultraviolet light
b. soil abrasion
c. a lack of adhesion
d. physical damage

13. Coating is inspected by machine and _____.

a. with meters
b. visually
c. ultrasonically
d. electrically

14. A holiday detector is also called a _____.

a. jeep
b. meter
c. cathode
d. primer detector

15. A fast way to locate holidays is the _____ survey.

a. line
b. magnetic
c. Pearson
d. X-ray

GLOSSARY

Trade Terms Introduced in This Module

Coating flaw: A gap or other flaw in pipe coating; it is also called a holiday.

Dielectric strength: The maximum electrical potential gradient that a material can withstand without rupture; it is usually specified in volts per millimeter of thickness.

Disbonding: A common coating failure in which the coating separates from the pipe.

Dope pots: Portable containers used to melt coal-tar enamel and maintain it at the required temperature for pipe coating.

Electroendosmosis: The use of electricity to cause a liquid to flow from an area of lesser concentration to one of greater concentration.

Enamel coating: A collective term for a variety of petroleum-based derivatives that are used to coat pipe.

Epoxy resin: A powdered resin coating that forms a skin over the steel when applied to the heated surface of pipe; it is usually applied at the mill.

Holiday: An imperfection in the pipe coating that exposes the unprotected surface to the environment.

Holiday detector: An electrical device used to locate a weak place in the coatings on pipelines and equipment; it is also called a jeep.

Pipe coating: A special material that is applied to the pipe to prevent water from coming into contact with the steel.

Pipe wrapping: A material applied on top of the pipe coating to protect the coating from damage.

Rock shield: A protective barrier placed around the pipe to shield it from damage while backfilling and during operation.

Tape coating: A protective coating of various materials that is wrapped around the pipe to protect it from corrosion.

CONTREN™ LEARNING SERIES — USER UPDATES

The NCCER makes every effort to keep these textbooks up-to-date and free of technical errors. We appreciate your help in this process. If you have an idea for improving this textbook, or if you find an error, a typographical mistake, or an inaccuracy in NCCER's Contren™ textbooks, please write us, using this form or a photocopy. Be sure to include the exact module number, page number, a detailed description, and the correction, if applicable. Your input will be brought to the attention of the Technical Review Committee. Thank you for your assistance.

Instructors – If you found that additional materials were necessary in order to teach this module effectively, please let us know so that we may include them in the Equipment/Materials list in the Instructor's Guide.

Write: Curriculum Revision and Development Department
National Center for Construction Education and Research
P.O. Box 141104, Gainesville, FL 32614-1104

Fax: 352-334-0932

E-mail: curriculum@nccer.org

Craft ______________________ Module Name ______________________

Copyright Date ______________ Module Number ______________ Page Number(s) ______________

Description

____

____

____

____

(Optional) Correction

____

____

____

(Optional) Your Name and Address

____

____

____

Module 62205-02

Pipe Inspection and Nondestructive Testing

(CT 38.1, 38.2, and 38.3)

Introduction to the Pipeline Operator Qualification Rule, API, and NCCER

It could be said that petroleum products keep America running. Petroleum products like oil and natural gas are the principal sources of energy in the United States and are used in many manufacturing and chemical processes. Petroleum products are primarily transported by pipeline. There are more than a million miles of liquid and natural gas pipeline in this country, ranging in diameter from two inches to four feet. These pipelines bring crude oil and natural gas from production fields to processing plants and refineries—and ultimately to the consumer.

Pipeline Operator Qualification

Petroleum products are considered hazardous, so pipeline personnel must take extra precautions and follow specific procedures when working with them. To help ensure the safe handling and transportation of petroleum products, the United States Department of Transportation (DOT) mandates safe practices as federal regulations.

One such regulation, the Operator Qualification (OQ) rule, requires pipeline operators to develop a written qualification plan to evaluate their personnel and their contract personnel (see *49 CFR Part 192 Subpart N—Gas*, and *49 CFR Part 195 Subpart G—Liquid*). The evaluation hinges on the operator's ability to perform certain covered tasks and to be able to recognize and respond to abnormal operating conditions that may be encountered while performing these covered tasks.

Covered Tasks

A pipeline performance task is deemed *covered* if the activity passes the following four-part test:

1. Is it performed on a pipeline facility?
2. Is it an operations or maintenance task?
3. Is it performed as a requirement of *49 CFR Part 195 (Liquid)* or *49 CFR Part 192 (Gas)*?
4. Does it affect the operation or integrity of the pipeline facility?

If the answer to all four questions is *yes*, then the activity is considered *covered* and must figure into the evaluation process of pipeline personnel.

American Petroleum Institute

The OQ rule and its related covered tasks provided the catalyst to develop a unified pipeline training and assessment program—so that pipeline operators and owners have ready access to the assessment and training tools they need. This effort was the direct result of the vision and leadership provided by The American Petroleum Institute (API) and API members.

Efforts to unify the industry are nothing new to API. Indeed, the impetus for forming the Institute in 1919 was the need to standardize engineering specifications for drilling and production equipment. Today, the most pressing issues revolve around public perceptions and government policies toward the industry. Speaking with one voice on these issues has become as essential as having interchangeable parts in the field. API's members determine the petroleum industry's positions on public policy as well as the standards that govern its day-to-day operations worldwide.

Consortium on Operator Qualification

API assembled a focus group whose primary purpose was to identify covered tasks and provide compliance recommendations to pipeline operators.

The Consortium on Operator Qualification (COOQ) became the guiding force for the identification and development of covered tasks for the liquid pipeline industry. Their work product, API publication 1161, *Guidance Document for the Qualification of Liquid Pipeline Personnel*, formed the basis for the resulting Pipeliner Training and Assessment Program.

The covered task list created in API publication 1161 has now evolved to include gas covered tasks and is maintained by NCCER. The most current covered task list can be downloaded at www.nccer.org.

National Center for Construction Education and Research

With the guidance and support of API, its members, and the Consortium on Operator Qualification, the National Center for Construction Education and Research (NCCER) was able to design and implement the Pipeliner Training and Assessment Program.

NCCER is a not-for-profit 501(c)(3) education foundation established in 1995 by the world's largest and most progressive construction and maintenance contractors and national contractor associations. NCCER was founded to address the

severe workforce shortage facing the construction and maintenance industries and to develop a standardized training process and curricula. Today, NCCER is supported by hundreds of leading construction and maintenance companies, manufacturers, and national associations and developed more than 35 construction and maintenance titles.

NCCER also maintains a National Registry, which contains the records of trainees and instructors who have received NCCER training. All completion information for every trainee is sent to the NCCER and filed within the Registry. The Registry can then confirm training and skills for workers as they move from company to company, state to state, or even within their own company.

Standardized Curriculum (NCCER's Contren™ Learning Series)

NCCER curricula are developed using the expertise of industry subject matter experts (SMEs). The use of SMEs as an integral part of every new development or revision ensures that the curriculum is current, industry-driven, and indicative of industry practices. The development and/or revision process is facilitated from start to finish by NCCER project managers, who work closely with the SMEs and a select team of technical writer(s) on the development and presentation of the curriculum.

Pipeline: Five Training Paths

The pipeline curriculum focuses on essential skills and competencies within five skill areas:

- Corrosion Control
- Electrical and Instrumentation
- Maintenance
- Mechanical
- Operations (Field and Control Center)

SMEs from the pipeline industry have determined the skills and competencies needed based on API's covered tasks and the consensus of the group/industry. Specific company information is generally not incorporated into the curriculum unless the SMEs determine that the information is the industry standard.

Training Modules and Covered Tasks

The smallest unit of instruction in the curriculum is a *module*. Each curriculum is composed of modules (the equivalent of stand-alone chapters) that are logically sequenced into complete *levels* of instruction. Every module is task-oriented and requires the successful completion of a written (knowledge-based) examination and a performance (hands-on) test. Modules may be individually selected and/or combined for efficient task training. Modules in the pipeline curricula, to the extent possible, will align with API-developed covered tasks. While not all modules will relate to covered tasks, all covered tasks will relate to at least one module.

Benefits of Training to the Pipeline Marketplace

Applying NCCER's standardized training process allows pipeline owners and contractors in the United States to establish the benchmark for acceptable skill and competency levels within the pipeline industry. This process will assist training organizations in developing and/or improving standardized pipeline training programs that effectively match their resources and operating conditions while responding to regulatory demands and to the needs of the industry. The process will also recognize training organizations that meet the minimum standards of competence, warranting regulatory, public, and professional confidence. This in turn will foster national unity within the pipeline industry for the mutual benefit of individual technicians, owners, contractors, and ultimately pipeline-users and society.

Additional Resources

To review the Pipeliner Training and Assessment Program Implementation and Resource Guide and the latest version of pipeline covered tasks, visit NCCER's Web site at www.nccer.org, or contact:

National Center for Construction Education and Research
PO Box 141104
Gainesville, FL 32614-1104
(888) 622-3720
Web site: www.nccer.org

For more information on API, please contact:

American Petroleum Institute
1220 L Street, Northwest
Washington, D.C. 20005
(202) 682-8000
Web site: www.api.org

To order *API Pub 1161*, contact Global Engineering Documents by calling 1-800-854-7149 or visiting their Web site at www.global.his.com.

Additional sources of information on the pipeline operator qualification regulation include:

DOT – Office of Pipeline Safety
www.ops.dot.gov

Transportation Safety Institute
www.tsi.dot.gov

COURSE MAP

This course map shows all of the modules in the second level of the Pipeline Maintenance curriculum. The suggested training order begins at the bottom and proceeds up. Skill levels increase as you advance on the course map. The local Training Program Sponsor may adjust the training order.

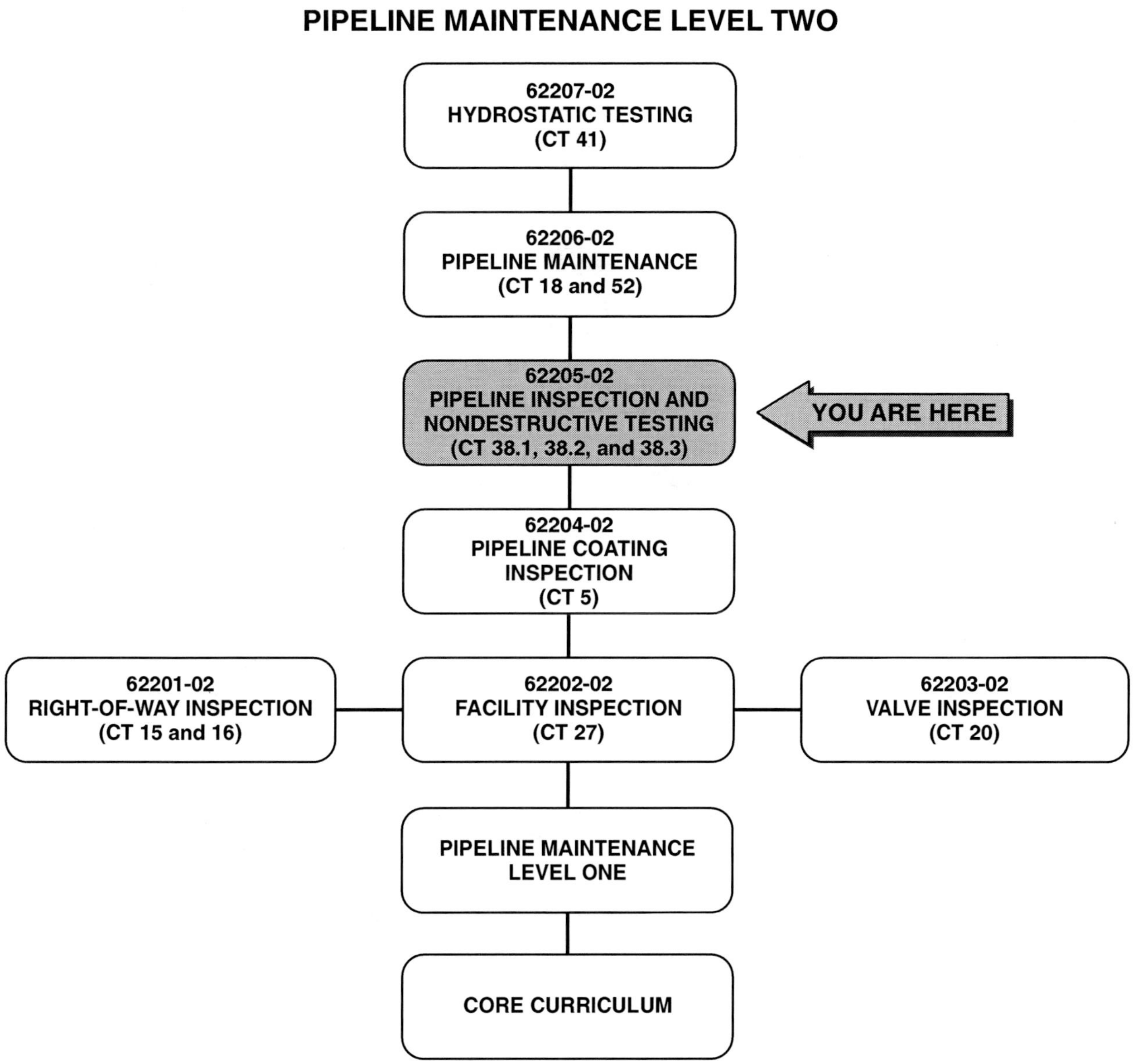

205CMAP.EPS

MODULE 62205-02 CONTENTS

Figures

Table

MODULE 62205-02

Pipe Inspection and Nondestructive Testing (CT 38.1, 38.2, and 38.3)

Objectives

When you have completed this module, you will be able to do the following:

1. Identify types of NDE testing.
2. Prepare welds for NDE testing.
3. Visually inspect pipe and pipe components (CT 38.1).
4. Direct/verify a nondestructive weld test (CT 38.2).
5. Visually inspect that welds meet DOT requirements (CT 38.3)

Prerequisites

Before you begin this module, it is recommended that you successfully complete the following: Core Curriculum; Pipeline Core, Modules 66101-02 and 66102-02; Pipeline Maintenance Level One; Pipeline Maintenance Level Two, Modules 62201-02 through 62204-02.

Required Trainee Materials

1. Pencil and paper
2. Appropriate personal protective equipment

1.0.0 ◆ INTRODUCTION

This module introduces nondestructive testing procedures. Once you are familiar with fabricating piping components, you are ready to perform NDE testing activities that test the integrity of pipe welds. One responsibility of a pipeline maintenance employee is to prepare welds for inspection and help technicians perform NDE testing. This module explains various methods of NDE testing.

2.0.0 ◆ PERFORM VISUAL INSPECTIONS

Visual inspections of pipe welds are the responsibility of the welder. Weld joints are visually inspected before the welding is started, as welding proceeds, and after the welding is completed to ensure that a quality weld is produced and that all code requirements are satisfied. If at any point a weld fails to pass inspection, it must be repaired or redone to the satisfaction of the inspection criteria.

2.1.0 Visually Inspect Pipe and Pipe Components

It is important to be able to visually inspect all pipe and/or components that are to be installed in the pipeline system to ensure that they are not damaged. Follow these steps to visually inspect pipe and components:

Step 1 Check components and consumables, including internal and external surfaces, upon receipt for:

- Corrosion
- Improper beveling
- Proper classification and condition of the welding rods
- Correct grade of pipe
- Roundness
- Buckling
- Surface defects, such as **cracks**, grooves, gouges, dents, and arc burns

Step 2 Clean the pipe to permit thorough inspection so that defects that might impair the pipe's strength may be detected.

Step 3 Check the hydrotesting paperwork for adherence to company specifications.

> **NOTE**
> This ensures that there are no defects in the materials and that the proper consumables and materials are being used. If the materials do not meet company specifications, document as required, and notify the proper personnel to verify the inconsistency.

Step 4 Visually inspect the pipe components. This visual inspection must ensure that:

- The welding is performed in accordance with the qualified welding procedures
- The weld is acceptable to API and DOT requirements as applicable
- The appropriate pipe supports or backfill material is in place to minimize stress on the pipe

> **NOTE**
> Visual inspection is any surveillance of all the phases of the operation, including surface preparation, line-up electrode selection, welding machine setting, preheat weather protection selection (if required), each welding pass, and completed weld inspection.

Step 5 Document the inspection on the appropriate company form, which should be signed by the inspector. Maintain permanent records detailing the grade, wall thickness, and manufacturer of the replacement pipe and its installation location.

> **NOTE**
> If the pipe component is of a different grade and/or wall thickness than the pipe being replaced, take care to ensure proper placement.

2.2.0 Perform Prefit-Up or Hold-Point Inspections

A hold point is an area that must be inspected before proceeding with the work activities, such as code welding. Perform the following procedures to perform a prefit-up inspection:

- Ensure that the base metal is clean and free of all foreign material.

> **NOTE**
> Foreign materials, such as chemicals, paint, grease, dirt, and scale, can combine with the weld metal and unfavorably affect the weld.

- Check the bevel angle on the pipe to ensure that it is within tolerance.
- Check the weld area to ensure that it is free of damage, such as nicks and gouges.

2.3.0 Visually Inspect Fit-Ups Using Hi-Lo Gauges

The following procedures explain how to perform a typical visual inspection of the fit-up. Always follow specific company policies.

Step 1 Check the pipe for corrosion, dents, and proper beveling.

Step 2 Check the classification and grade of the consumables and the pipe.

Step 3 Locate and review the hydrotesting paperwork for all materials.

Step 4 Check the fit-up for misalignment or mismatch with a Hi-Lo gauge. *Figure 1* shows measuring a mismatch with a Hi-Lo gauge.

Step 5 Check the fit-up for the proper root gap with a taper gauge. *Figure 2* shows measuring a root gap with a taper gauge.

Step 6 Inspect the tack welds for any weld **defects**.

> **NOTE**
> Tack welds can exhibit defects such as cracks, crater cracks, inclusions, poor fusion, undercut, inadequate penetration, and burn-through. The starting and stopping ends of the tack weld should be ground so that they can be incorporated into the root pass without developing defects.

Step 7 Document the inspection results in the project file/log.

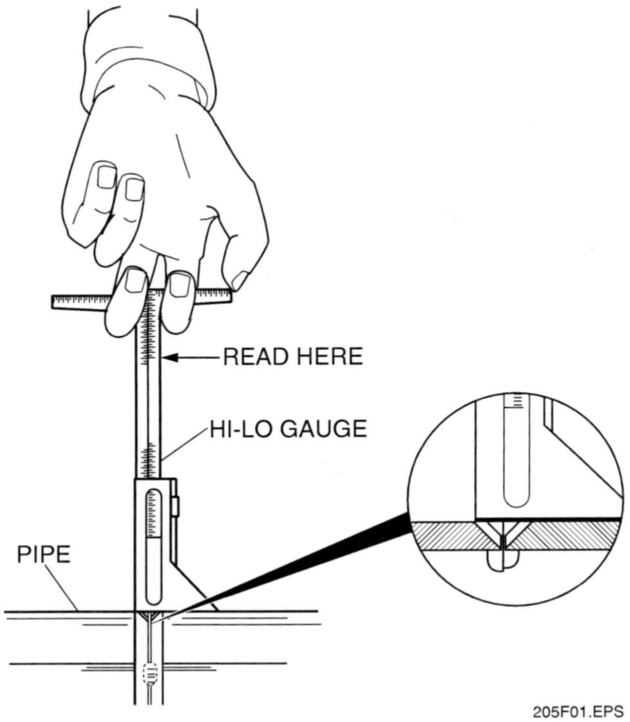

Figure 1 ◈ Measuring a mismatch with a Hi-Lo gauge.

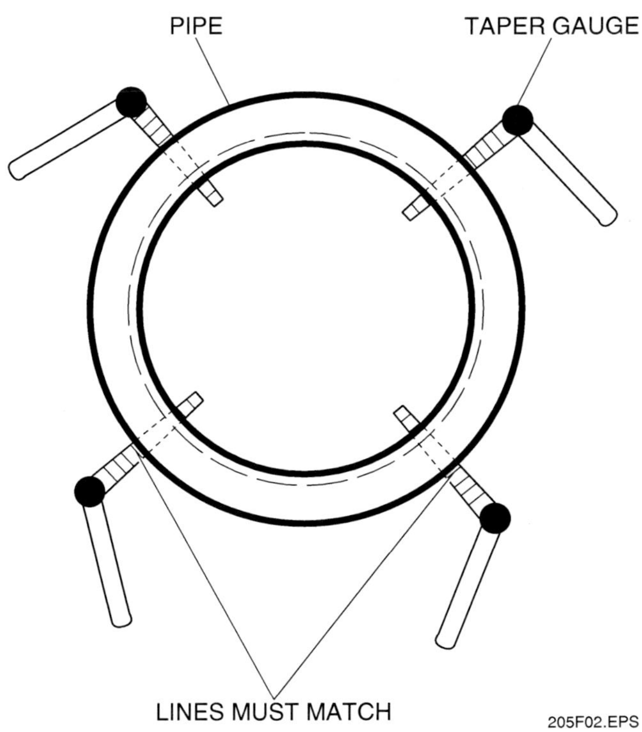

Figure 2 ◈ Measuring root gap with a taper gauge.

2.4.0 Inspect Root Passes

Each root pass of a weld must be thoroughly inspected to identify any defects. All defects in the root pass must be repaired before welding is resumed because defects, especially cracks, in the root pass can crack out or propagate through the entire finished weld. The weld should be inspected on the inside and outside if possible.

The following explains how to inspect a typical root pass. Always follow specific company policies. When inspecting a typical root pass:

- Inspect the root pass weld bead for longitudinal and transverse cracks.
- Inspect the **craters** and the heat-affected base metal on both sides of the weld for cracks.
- Inspect the weld bead for **slag inclusions**.
- Inspect the seams where the weld metal joins the base metal and the starting/stopping points for lack of fusion.
- Look for any excessively convex or concave surfaces, which indicate improper penetration, on the inside of the weld bead.
- Examine the entire surface of the root pass for holes or voids, which indicate burn-through.
- Look for grooves in the base metal, which indicate **undercut**, along both sides of the weld bead.
- Inspect the entire weld bead closely for one or more tiny pinholes, which indicate **porosity**.
- Inspect the entire weld area to ensure that all slag, spatter, or other substances that could inhibit the fusing of subsequent weld beads to the root bead have been removed.

If any unacceptable conditions are found during the root pass inspection, reject the root pass, and ensure that all defects are repaired before the welding is resumed.

The procedure for inspecting interpass welds is the same as that for inspecting root pass welds. Each interpass weld should be inspected thoroughly for cleanliness or for any defects before the next weld pass is made. Any defects must be repaired before welding is resumed.

2.5.0 Visually Inspect Completed Welds

After the welding has been completed, the weld must be visually inspected thoroughly to ensure that there are no unacceptable defects and that a quality weld has been produced.

The following explains how to inspect the completed weld. Always follow specific company policies. When inspecting a completed weld:

- Inspect the entire weld to ensure that it is free of slag and scale that could mask other defects.
- Inspect the entire weld to ensure that there are no cracks or inclusions.
- Ensure that the amount of porosity in the weld is within specified limits.
- Ensure that the tie-in of the weld metal to the base metal is smooth.
- Ensure that undercut, if any, is within specified limits and does not violate the minimum wall thickness of the base material.
- Ensure that all weld spatter and arc strikes have been removed.
- Inspect the heat-affected zone for any defects.
- Ensure that all weld beads have complete penetration and fusion for the full length of the weld.
- Ensure that the crown of the weld meets specifications.
- Inspect the entire weld to ensure that it has no craters.

NOTE
All craters must be filled to the full cross section of the welds.

Any defects found in the weld must be repaired. Some defects, such as spatter, arc strikes, porosity, and inclusions, may require only a small amount of grinding to repair them. Repair of other more serious defects, such as cracks, may require completely reworking the weld.

2.6.0 Inspect Completed Welds Using Borescopes

Borescopes are used for visual inspection to detect surface flaws inside pipe, valves, and rotating equipment in areas that are not restricted from direct visual inspection. *Figure 3* shows three typical types of borescopes.

The flexible borescopes are most commonly used by pipeline workers. Two of the most common types of flexible borescopes are flexible fiberscopes and videoscopes with charge coupled device (CCD) probes.

2.6.1 Flexible Fiberscopes

A fiberscope consists of a light guide bundle, an image guide bundle, an objective lens, interchangeable viewing heads, and remote controls to control the rotation of the tip. Fiberscopes are available in varying diameters and come in lengths up to 40 feet. The interchangeable viewing heads provide various directions and fields of view on a single fiberscope. Fields of view are typically 40 to 60 degrees; special applications can range from 10 to 120 degrees. Most fiberscopes provide adjustable focusing of the objective lens.

Specialized training is required to use a flexible fiberscope.

2.6.2 Videoscopes with CCD Probes

Videoscopes with CCD probes transmit color or black-and-white images to a video monitor. The end of the electronic videoscope contains a CCD chip, which consists of thousands of light-sensitive elements in a pattern of rows and columns. Videoscopes with CCD probes produce images similar to borescopes, but they normally produce a higher resolution. The most significant difference is that videoscopes with CCD probes have a longer working length than fiberscopes. Videoscopes with CCD probes are operated in the same manner as flexible fiberscopes. *Figure 4* shows a typical videoscope.

3.0.0 ◆ NONDESTRUCTIVE EXAMINATION AND EVALUATION

Nondestructive examination (NDE) methods are used to test equipment and components without damaging them. Nondestructive examination consists of one or more types of tests that quantify defects or detect future failures. With NDE, it is possible to detect defects or changes that will lead to premature failure. The results can be used to find the real cause of the change or abnormality and correct the problem, not just the symptom. *Table 1* lists the types of NDE. *Figure 5* shows weld discontinuities.

Safety should be the first concern when preparing equipment for NDE testing. Most equipment to be tested must be cleaned using a grinder with a flapper wheel, wire brush, or 14-inch half-round bastard file. The following should be performed when preparing for NDE testing activities. Always follow specific company policies. To prepare for NDE testing:

- Have the safety engineer check the area to ensure that adequate ventilation is available when using chemicals to perform NDE testing.
- Inspect grinders for broken, missing, or loose parts.

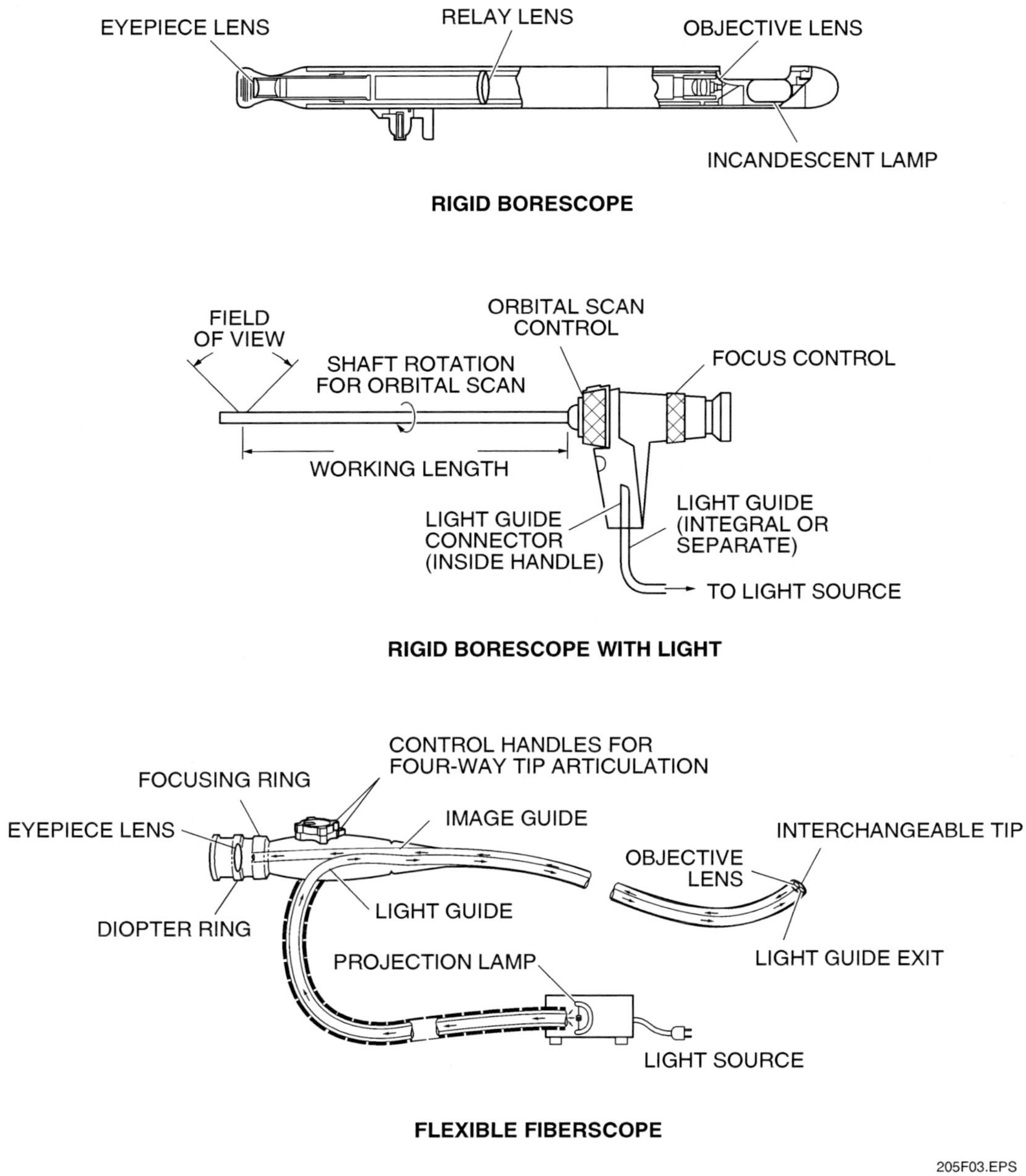

Figure 3 ◆ Three typical types of borescopes.

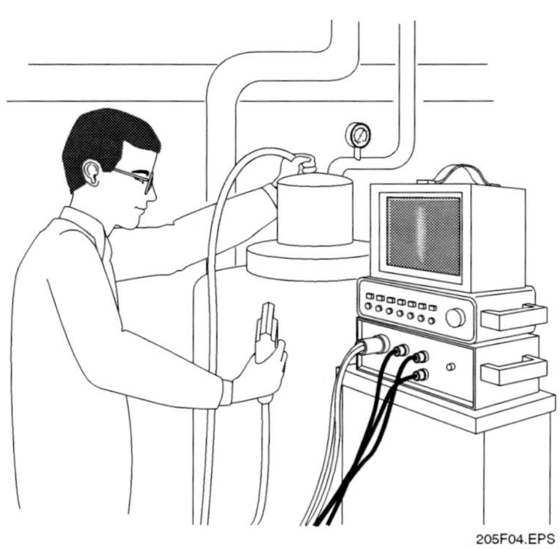

Figure 4 ◆ Typical videoscope.

- Check the grinding wheel to ensure that the revolutions per minute (rpm) rating is higher than that of the grinder it is to be used on.
- Barricade the work area to prevent through-traffic.
- Take special precautions when preparing to perform **radiographic testing** to prevent over-exposure to radiation.
- Inspect the scaffolding and work platform for safe access and egress.

As a pipeline worker, you will not be expected to perform NDE testing unless you are prepared to do so through specialized training and certification. Your main duty is to prepare the weld for inspection and assist the technician who is performing the test.

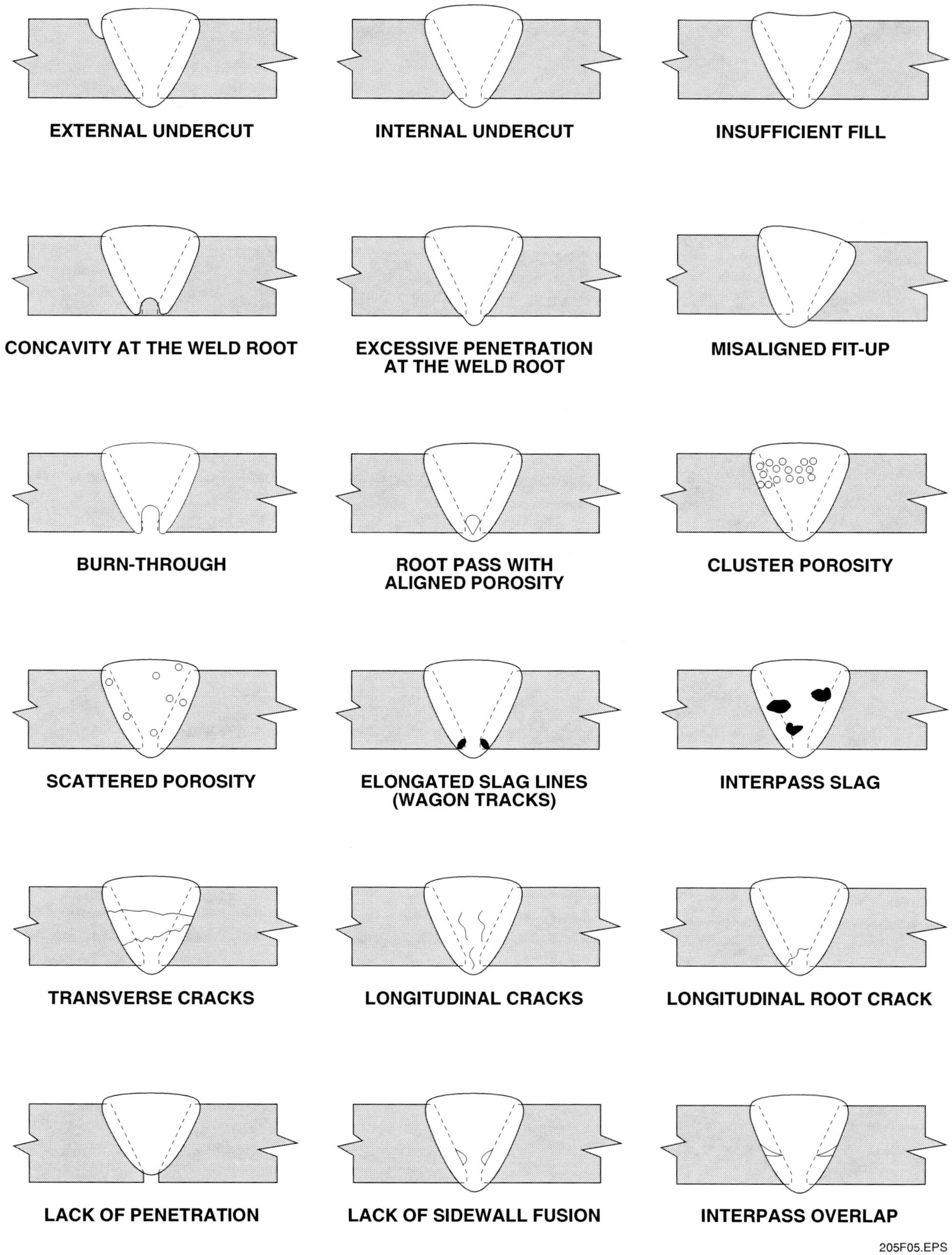

Figure 5 ◆ Weld discontinuities.

Table 1 Types of NDE

NDE	Type of Reading	Advantages	Limitations	Examples of Uses
Liquid penetrant	Detects surface openings due to cracks, porosity, seams, or folds	Is inexpensive, easy to use, readily portable, and sensitive to small surface flaws	Flaw must be open to the surface; is not useful for porous materials	Fillet welds
Magnetic particles	Detects leakage of magnetic flux caused by surface or near-surface cracks, voids, inclusions, and material or geometry changes	Is inexpensive and sensitive to both surface and near-surface flaws	Is limited to ferromagnetic material; possibly requires surface preparation and postinspection demagnetization	Fillet welds
Ultrasonics	Detects changes in acoustic impedance caused by cracks, nonbonds, inclusions, or interfaces	Can penetrate thick materials, is excellent for crack detection, and can be automated	Normally requires coupling either by contact to surface or immersion in a fluid; problems possibly presented by orientation, detection, or interpretation of the defect	Girth weld thickness and pipeline thickness
Radiography	Detects changes in material density from voids, inclusions, material variations, and placement of internal parts	Can be used to inspect a wide range of materials and thicknesses, is versatile, and provides a record of inspection on film	Requires expensive radiation safety precautions; doesn't easily detect cracks	Pipeline welds for penetration, inclusions, and voids; verification of parts in assemblies
Visual-optical	Detects surface characteristics, such as finish, scratches, cracks, or color, and strain in transparent material	Is often convenient and can be automated	Can be applied only to surfaces, through surface openings, or to transparent material	Metal for surface finish and uniformity
Eddy currents	Detects changes in electrical conductivity or magnetic permeability caused by material variations, cracks, voids, or inclusions	Is readily automated and moderately priced	Is limited to electrically conducting materials and has limited penetration depth; interpretation of defect signals possibly difficult	Heat exchanger tubes for wall thinning and cracks; verification of material heat treatment

205T01.EPS

3.1.0 Liquid Penetrant Test

Liquid penetrant inspection is a method of highlighting cracks and gaps that are not visible to the naked eye in the surface of smooth or nonporous samples. The liquid penetrant seeps into surface cracks, porosity, delaminations, or shrinkage areas by capillary action. It is also used to highlight wear patterns and measure stress cracking.

Liquid penetrant inspection involves wetting the surface of a test object so that the liquid flows over that surface to form a continuous and uniform coating and migrates into cracks and surface defects. The liquid is then washed off the surface, and a developer is placed on the surface. As the developer draws the penetrant out of the defects, it is stained by the penetrant and highlights the cracks and surface irregularities to form a visual image of the flaw.

With liquid dye penetrants, two types of indications are revealed: true indications and false indications. True indications are caused by the penetrant bleeding out from actual **discontinuities** in the metal. The standard defects indicated are fatigue cracks, **underbead cracks**, pits, and porosities. Large fatigue cracks are indicated by wide lines that become apparent quickly after the developer is applied. Underbead cracks are undersurface cracks that bleed through the surface and are indicated by a line of dots that appear a few minutes after the developer is applied. Porosity is indicated by dots that come to the surface almost immediately.

The major causes of false indications are failure to apply the penetrant correctly or rough, irregular surfaces of the test metal. The procedure for liquid dye penetrant testing must be carefully performed in order to achieve the proper results.

3.2.0 Magnetic Particle Test

When a magnetic field is generated in and around a part made of a ferromagnetic material, such as steel or iron alloys, and the lines of magnetic flux are intersected by a defect such as a crack, magnetic poles are generated on either side of the defect. Magnetic particle inspection works only for materials that can be magnetized, such as steel and nickel alloys.

In a magnetic particle test, the sample is magnetized by an electrical current, and magnetic particles are applied over the surface. These particles can be applied either dry or in a liquid carrier, such as oil or water. The flaws on or just under the surface of the sample modify the magnetic field so that the particles outline the defect in a pattern and highlight the shape and location of the flaw.

This method is very sensitive to minor surface cracks and flaws that are near the surface. The disadvantages are the limitation to magnetic samples and the inability to read deeper flaws in the sample.

Maximum sensitivity is obtained when the part is oriented perpendicular to the magnetic field, and the strength of the field saturates the entire part. The magnetic particle equipment uses either AC or DC current and can be applied to the part with coils, probes, or flexible conductors that can wrap around a test piece.

Due to the variety of shapes of samples, the inductor for the test must come in a variety of sizes and shapes. Some sample parts, such as gears, balls, and ring-shaped parts, can be placed in the magnetic field without contact from a probe. Other samples use coils, clamps, central conductors, or other forms of direct contact with the sample. A yoke is used on a welded joint to induce a magnetic field around the weld seam and show any discontinuities as extra particles scattered outside the weld seam. *Figure 6* shows electromagnetic yokes.

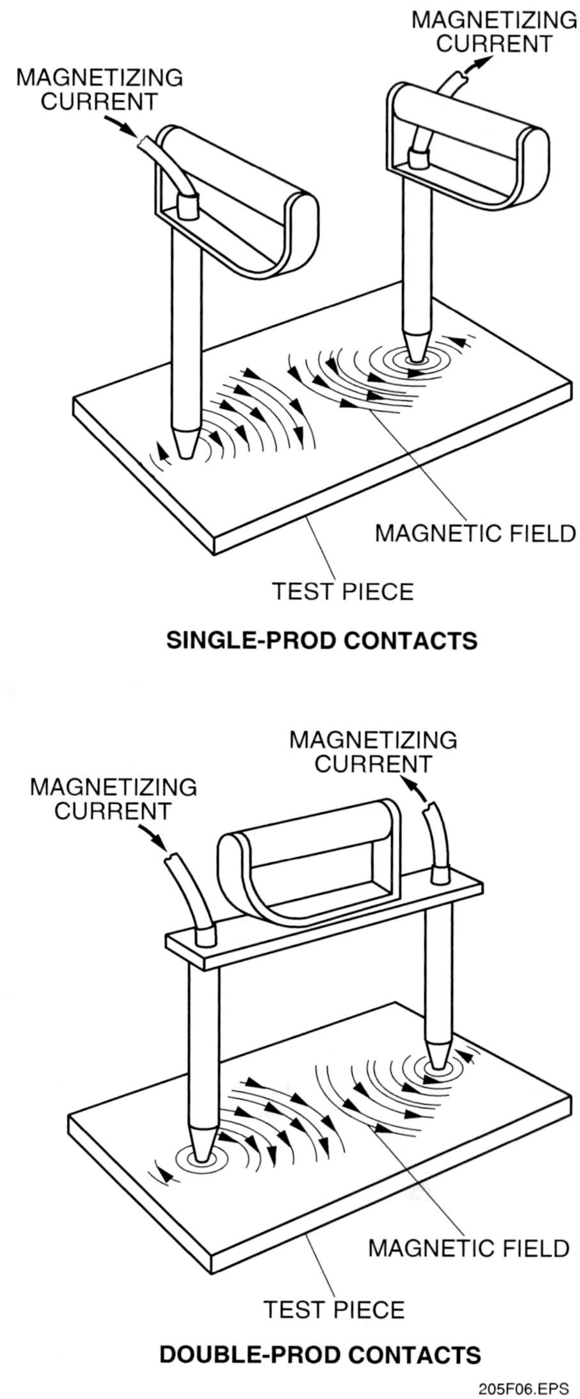

Figure 6 ◆ Electromagnetic yokes.

3.3.0 Ultrasonic Test

Ultrasonics use high-frequency sound waves to detect the noise generated by plant equipment. Ultrasonic instruments measure thickness in materials or "listen" for turbulence in pipes to detect internal leaks and malfunctions. Other ultrasonic test instruments send focused, amplified sound waves into solid materials and compare them with the return wave to find flaws, such as corrosion, bad welds, or casting flaws, below the surface. The frequency range for ultrasonic monitoring is from 20,000 Hz to 100 kiloHz.

In many ways, a beam of ultrasound is similar to a beam of light. Both are waves and travel in a constant velocity in a consistent medium. The velocity depends on the medium, not on the properties of the wave form. Like a beam of light, ultrasonic waves are bent, reflected from surfaces, refracted when they cross boundaries of a different surface, and diffracted at edges or around obstacles. This beam, when reflected back, creates a signature that is interpreted as an image by the ultrasonic monitor.

Ultrasonic monitors can penetrate deep and with high sensitivity and accuracy. The output is readily digitized, and operation of the monitor is entirely nonhazardous. Some of the different types of ultrasonic techniques are pulse-echo, pulse-transmission, ultrasonic attenuation, continuous wave resonance, and ultrasonic spectroscopy. Ultrasonic monitors are very portable but require experienced technicians to monitor equipment. Reading an ultrasonic flaw detector requires some practice because the screen image shows the peaks of back reflection. *Figure 7* shows an ultrasonic flaw detector and an angle beam transducer.

3.4.0 Radiography

Radiography is a nondestructive examination method that uses a beam of radiation that penetrates the **weldment**. Radiation is very sensitive and can be used to inspect the density of almost any ferrous, nonferrous, organic, or inorganic material. Some of the radiation energy is absorbed, and the intensity of the beam is reduced. The variations in the beam intensity are recorded on film. These variations are seen as differences in shading that are typical of the types and sizes of any discontinuities present.

Radiography can detect both surface and subsurface discontinuities. Because of the extremely powerful nature of **gamma** rays, much radiographic work is now done with iridium sources. The principal advantages of gamma ray inspection are the low cost and portability of the source, although special precautions must be taken when working with iridium materials to protect personnel from overexposure to radiation.

> **WARNING!**
> Only certified technicians are permitted to perform radiographic activities. Never cross a radiation barrier.

The radiographic method is used to detect internal defects and to check for proper alignment of assembled parts. Radiation is also used extensively to examine welds, check wall thickness, and examine the quality of casting and forging. *Figure 8* shows a radiographic test using gamma rays. Refer back to *Figure 5* for examples of surface and subsurface discontinuities.

3.4.1 Read Surface Discontinuities

The following are types of surface discontinuities that can be detected by radiographic testing:

- *External undercut* – External undercut is a gouging out of the piece to be welded alongside the edge of the top or external surface of the weld. The film shows an irregular, darker density along the edge of the weld image.
- *Internal undercut* – Internal, or root, undercut is a gouging out of the piece to be welded alongside the edge of the bottom or internal surface of the weld. The film shows an irregular, darker density near the center of the width of the weld and along the edge of the root pass.
- *Insufficient fill* – Longitudinal grooves indicate insufficient fill. The film shows a weld density darker than the density of the pieces being welded and extending across the full width of the weld image.
- ***Concavity** at the **weld root*** – The film shows an elongated, irregular, darker density with fuzzy edges in the center of the width of the weld image.
- *Excessive penetration at the weld root* – The film shows a lighter density in the center of the width of the weld image with a straight, darker density in the center of the width of the weld image, either extended along the weld or in isolated, circular drops.
- *Weld with offset or mismatch with lack of penetration of the pieces to be welded with insufficient filling of the bottom of the weld* – The film shows an abrupt density change across the width of the weld image with a straight, darker density line at the center of the width of the weld image along the edge of the density change.

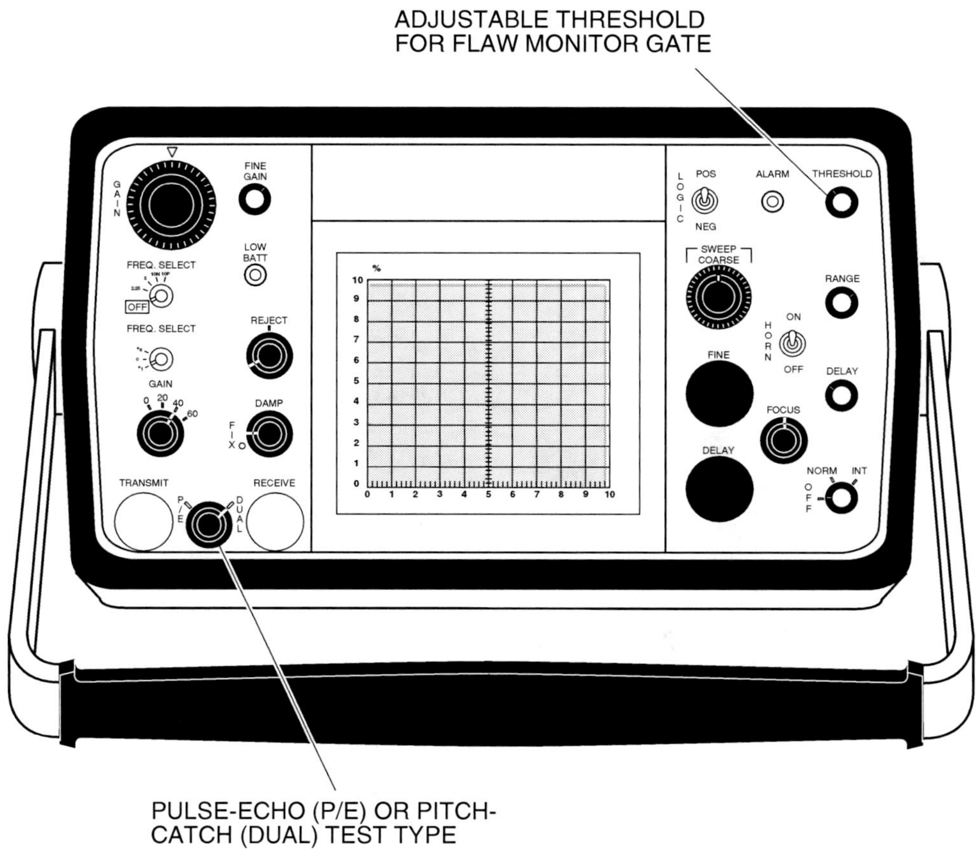

ULTRASONIC FLAW DETECTOR

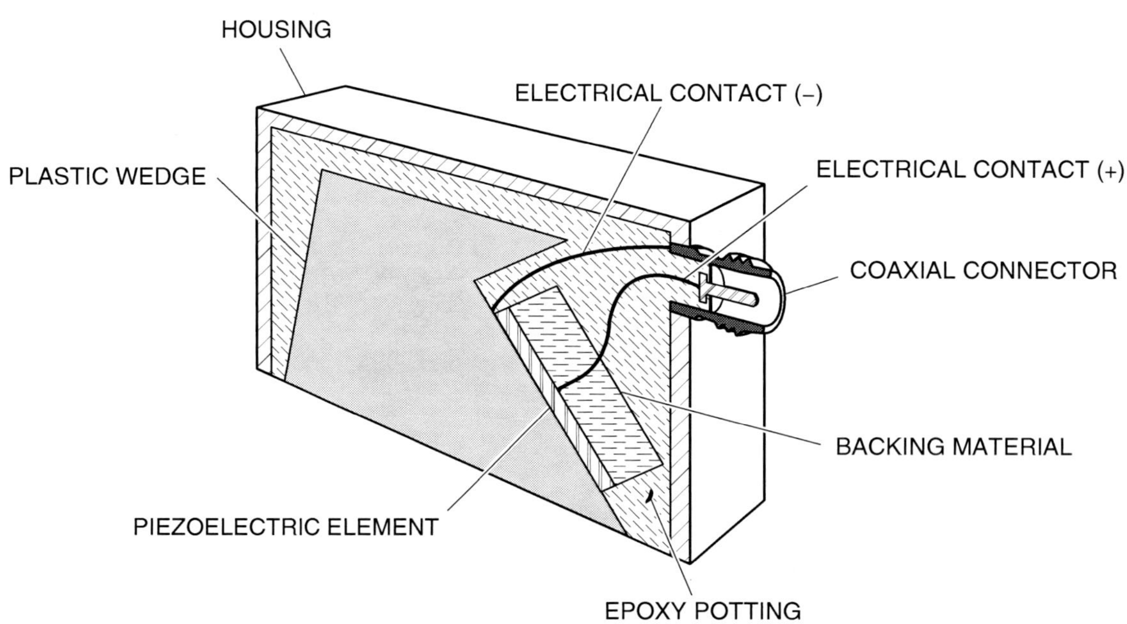

ANGLE BEAM TRANSDUCER

205F07.EPS

Figure 7 ◆ Ultrasonic flaw detector and an angle beam transducer.

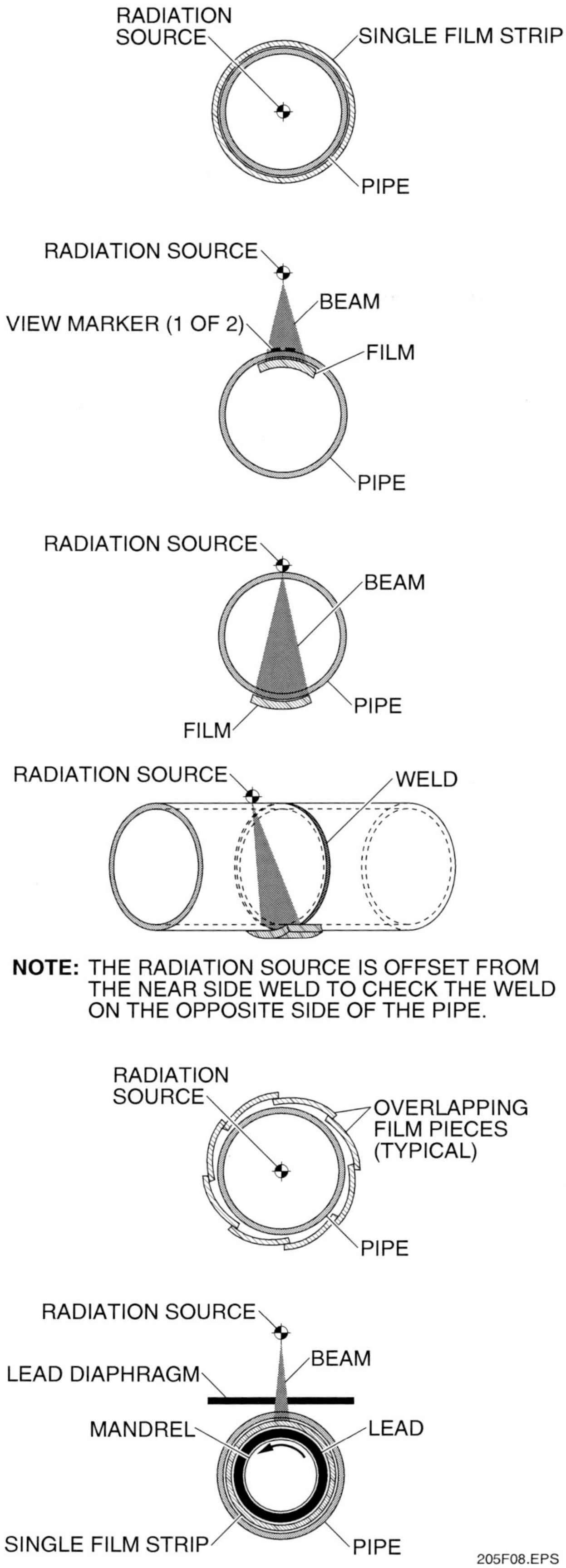

Figure 8 ◆ Radiographic test using gamma rays.

- *Misaligned fit-up* – Misaligned fit-up is a weld with offset or a mismatch of the pieces to be welded. The film shows an abrupt change in film density across the width of the weld image.
- *Burn-through* – A burn-through is a severe depression or a crater hole at the bottom of the weld. The film shows a localized, darker density with fuzzy edges in the center of the width of the weld image, which could appear wider than the root pass image.

3.4.2 Read Subsurface Discontinuities

The following are types of subsurface discontinuities that can be detected by radiographic testing:

- *Root pass with aligned porosity* – Involves rounded and elongated voids in the bottom of the weld along the center line of the weld. The film shows rounded and elongated darker density spots, which may be connected in a straight line in the center of the weld image.
- *Cluster porosity* – Involves rounded or slightly elongated voids grouped together. The film shows rounded or slightly elongated darker density spots in clusters randomly spaced.
- *Scattered porosity* – Involves rounded voids that are random in size and location. The film shows rounded spots of darker densities that are random in size and location.
- *Elongated slag lines (wagon tracks)* – Impurities that solidified on the surface after welding and were not removed between passes. The film shows elongated, parallel, or single darker density lines irregular in width and slightly winding in the lengthwise direction.
- *Interpass slag inclusions* – Usually nonmetallic impurities that solidified on the weld surface and were not removed between weld passes. The film shows an irregularly shaped darker density spot that is usually slightly elongated and randomly spaced.
- *Transverse cracks* – A fracture in the weld metal running across the weld. The film shows feathery, twisting lines of darker density running across the width of the weld image.
- *Longitudinal cracks* – A fracture in the weld metal running lengthwise in the welding direction. The film shows feathery, twisting lines of darker density running lengthwise along the weld at any location in the width of the weld image.
- *Longitudinal root cracks* – A fracture in the weld metal at the edge of the root pass. The film shows feathery, twisting lines of darker density along the edge of the image of the root pass.

The twisting feature helps to distinguish the root crack from incomplete **root penetration**.

- *Lack of penetration* – Occurs when the edges of the pieces have not been welded together, usually at the bottom of a single V-groove weld. The film shows a darker density band with very straight, parallel edges in the center of the weld image.
- *Lack of sidewall fusion* – Involves elongated voids between the weld beads and the joint surfaces to be welded. The film shows elongated, parallel, or single darker density lines that sometimes have darker density spots along the lack of fusion lines, which are straight in the lengthwise direction of the weld, not winding like elongated slag lines. One edge of the lack of fusion line may be very straight as with the lack of penetration. The lack of sidewall fusion images will not be in the center of the weld image.
- *Interpass overlap* – Involves lack of fusion areas along the top surface and edge of lower passes. The film shows small spots of darker densities, some with slightly elongated tails that are aligned in the welding direction and not in the center of the weld image.
- *Tungsten inclusions* – Random bits of tungsten fused into, but not melted into, the weld metal. The film shows irregularly shaped, low-density spots that are randomly located in the weld image.

3.5.0 Conductivity Tests

Pipeline workers are sometimes called upon to perform conductivity tests, or holiday tests, on an underground piping system. Conductivity testing ensures that there is no unprotected piping that is not covered with coating. Conductivity tests use voltage to test the conductivity of a material. The **holiday testers** have variable voltage ratings, and the voltage rating used must comply with the specifications of the piping system. *Figure 9* shows a holiday tester.

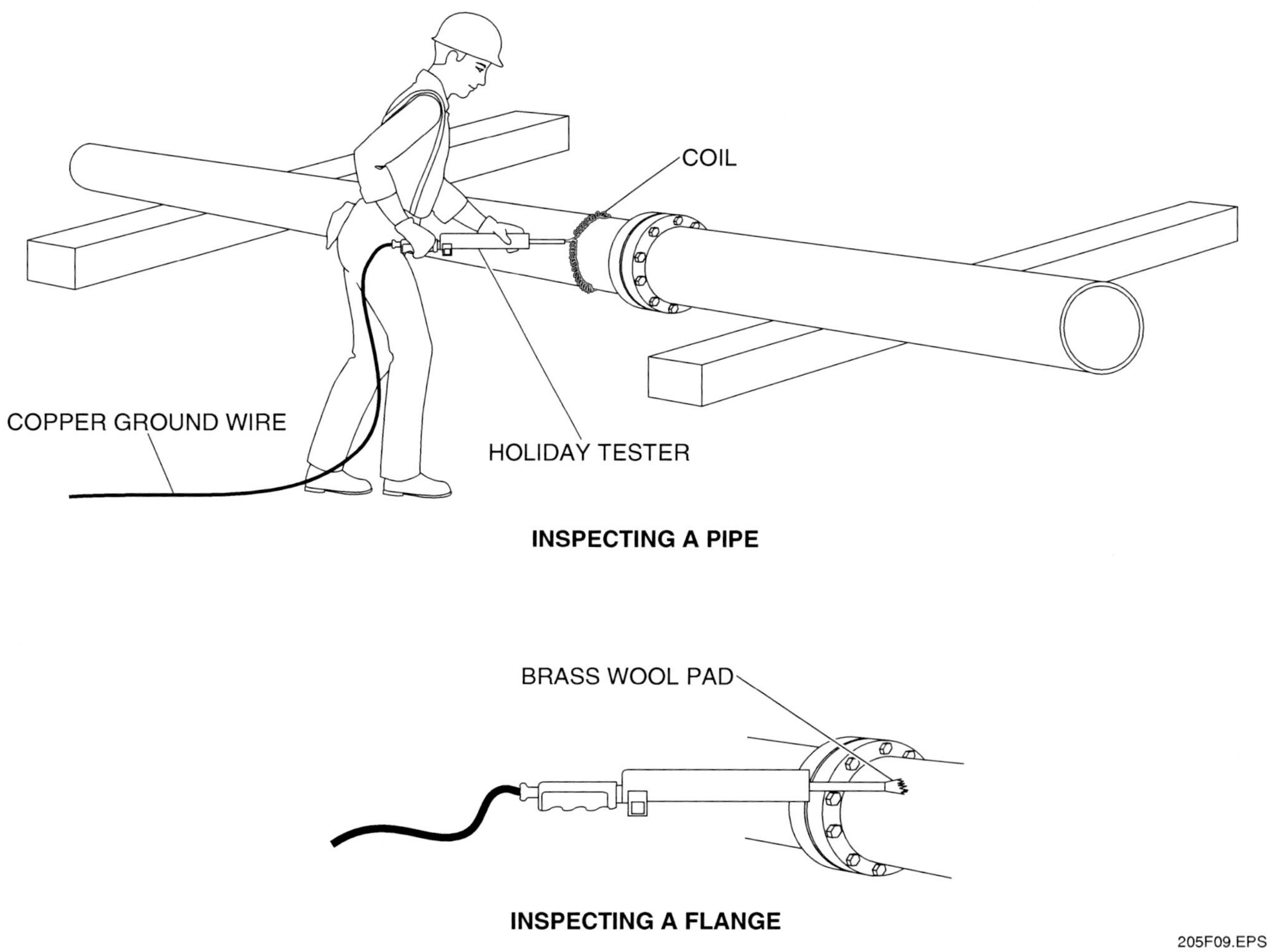

Figure 9 ◈ Holiday tester.

4.0.0 ◆ DIRECT AND VERIFY NONDESTRUCTIVE WELD TEST

Pipeline personnel must be able to determine what type of nondestructive test (NDT) is to be used to test the integrity of a pipe, weld, or component. Follow these steps to direct and verify an NDT:

Step 1 Determine what NDT process to use for each component or weld.

Step 2 Determine that the technicians are certified in the NDT process that will be used.

> **NOTE**
> This is to ensure that qualified personnel interpreting the process and data will provide satisfactory results.

Step 3 Observe that the technician is conducting the appropriate test (mag particle, dye penetrate, ultrasonic, gamma ray) for each component or weld. Check the results for abnormal conditions or defects.

> **NOTE**
> If any repairs or cutouts of defects need to be made, proceed to Step 4. Otherwise, proceed to Step 6. If any defects are found, then ensure that the location and nature of the defect are recorded in project file and repaired immediately.

Step 4 Repair or cut out the defect, and retest for weld integrity.

Step 5 Ensure the weld integrity before backfilling.

Step 6 Record all required data to update the piping records.

Step 7 Identify potential abnormal operating conditions and the appropriate action to be taken.

5.0.0 ◆ VISUALLY INSPECT THAT THE WELD MEETS DOT REQUIREMENTS

Pipeline personnel must be able to verify that welds meet DOT regulations for nondestructive testing, **X-raying** pipeline welds, and identifying any defects that may affect the integrity of the pipeline tie-in or component replacement. Follow these steps to visually inspect that the weld meets DOT requirements:

Step 1 Determine which weld inspection process is appropriate for the weld.

Step 2 Ensure that the NDT technicians are certified for the inspection process on the component or type of weld.

> **NOTE**
> This is to ensure that qualified personnel interpreting the process and data are certified for the appropriate process for inspecting the various types of welds.

Step 3 Observe that the technician is conducting the appropriate test (mag particle, dye penetrate, ultrasonic, gamma ray) for each component or weld. Check the results for abnormal conditions or defects.

> **NOTE**
> If any repairs or cutouts of defects need to be made, proceed to Step 4. Otherwise, proceed to Step 6. If any defects are found, then ensure that the location and nature of the defect are recorded in project file and repaired immediately.

Step 4 Repair or cut out the defect, and retest for weld integrity.

Step 5 Ensure weld integrity before backfilling.

Step 6 Record all required data.

Step 7 Identify potential abnormal operating conditions and the appropriate action to be taken.

6.0.0 ◆ ABNORMAL OPERATING CONDITIONS

Abnormal operating conditions (AOCs) may be encountered when inspecting pipe, pipe components, and welds, as well as when directing and verifying nondestructive weld tests. It is important to react correctly and promptly to these conditions. AOCs that may be encountered when performing these tasks include:

- *Component failure* – If a valve, weld/seam, flange, or joint fails, document the failure as required, and notify the proper personnel to correct the failure.

- *Fire/explosion* – If a hazardous liquid or gas is ignited, evacuate the area immediately, and notify local emergency personnel.
- *Pipeline system damage* – If a pipeline or pipeline segment is wrinkled, buckled, over-stressed, gouged, or dented, document the findings as required, and notify the appropriate personnel to correct the damage.
- *Unexpected hazardous liquid or carbon dioxide* – If an isolation device fails, causing a release of a hazardous liquid or carbon dioxide, immediately eliminate ignition sources, and notify the appropriate personnel.

Summary

To perform NDE testing activities, you must be proficient in preparing welds, valves, piping, and vessels for the appropriate NDE test. This module explains the different types of NDE testing that can be performed and the equipment used to perform the testing.

Procedures for preparing a weld for inspection and performing the actual inspection are also covered. The types of tests that are discussed in this module are the dye penetrant, magnetic particle, ultrasonic, and radiographic tests.

Review Questions

1. A gauge used to measure pipe mismatch during installation is a _____ gauge.

a. taper
b. gap
c. Hi-Lo
d. bevel

2. A gauge used to measure root gap is a _____ gauge.

a. taper
b. gap
c. Hi-Lo
d. bevel

3. The ends of tack welds must be _____ before welding over them.

a. sandblasted
b. ground down
c. chipped off
d. hammered

4. Holes or voids in the root pass indicate _____.

a. improper penetration
b. burn-through
c. lack of fusion
d. undercut

5. When welding, all craters must be filled to the _____ of the weld.

a. full cross section
b. full depth
c. bottom
d. center

6. The borescope most commonly used by pipeline workers is the _____ type.

a. rigid
b. rigid with light
c. stationary
d. flexible

7. _____ is used to high-light wear patterns.

a. Magnetic particle testing
b. Ultrasonic testing
c. Liquid penetrant testing
d. Radiography

8. Ultrasonic testing uses high-frequency _____ waves to detect noises.

a. light
b. magnetic
c. vibration
d. sound

9. Radiography can detect both surface and _____ defects.

a. subsurface
b. dye penetrant
c. shading
d. radiation

10. Longitudinal grooves in welds indicate _____.

a. craters
b. cracks
c. insufficient fill
d. bad root passes

11. Interpass slag inclusions are usually _____ impurities.

a. metallic
b. nonmetallic
c. ferrous
d. nonferrous

12. A fracture in the weld metal running _____ is a transverse crack.

a. beside the weld
b. lengthwise in the weld direction
c. through the root pass
d. across the weld

13. Conductivity tests are also called _____ tests.

a. direct
b. dye penetrant
c. holiday
d. vacuum

14. For any defect that is found, the _____ must be recorded.

a. depth
b. width
c. thickness
d. location

15. Verification of welds must meet _____ regulations.

a. OSHA
b. DOT
c. API
d. AWS

GLOSSARY

Trade Terms Introduced in This Module

Arc burn: The result of momentarily touching an electrode holder or ground clamp to the pipe surface.

Concavity: The maximum distance from the face of a concave fillet weld perpendicular to a line joining the toes of the weld.

Crack: A fracture-type discontinuity characterized by a sharp tip and high ratio of length and width to the opening displacement.

Crater: A depression at the termination point of a weld bead.

Defects: Discontinuities that render a part or product unable to meet minimum applicable acceptance standards or specifications.

Discontinuity: An interruption or gap in the typical structure of a weldment, such as a lack of consistency in the mechanical, metallurgical, or physical characteristics of the base material or weldment.

Gamma: A high-energy photon emitted by a nucleus that is in transition between two energy levels.

Holiday tester: An electrical device that is used to locate a gap or void in the corrosion protection coating of a metal surface.

NDE: Nondestructive examination of materials.

Porosity: Cavity-type discontinuities that are formed by gas entrapped in the weld during solidification.

Radiographic testing: An NDE testing method that produces a photographic image of an opaque specimen by transmitting a beam through it onto an adjacent photographic film.

Root penetration: The depth that a weld extends into the root of a joint measured on the center line of the root cross section.

Slag inclusions: Particles of impurities that are retained in the weld metal after the metal solidifies and that can occur anywhere in the weld bead.

Underbead crack: A crack in the heat-affected zone of the base metal that generally does not extend to the surface of the base metal.

Undercut: A groove melted into the base metal adjacent to the toe or root of a weld and left unfilled by weld metal.

Weld root: The points at which the back of the weld intersects with the base metal surfaces.

Weldment: An assembly in which the component parts are joined by welding.

X-ray: To penetrate with electromagnetic radiation.

Additional Resources

This module is intended to present thorough resources for task training. The following reference is suggested for further study. This is optional material for continued education rather than for task training.

ASTM Handbook, Volume 17, Nondestructive Evaluation and Quality Control, 1992. ASM International.

CONTREN™ LEARNING SERIES — USER UPDATES

The NCCER makes every effort to keep these textbooks up-to-date and free of technical errors. We appreciate your help in this process. If you have an idea for improving this textbook, or if you find an error, a typographical mistake, or an inaccuracy in NCCER's Contren™ textbooks, please write us, using this form or a photocopy. Be sure to include the exact module number, page number, a detailed description, and the correction, if applicable. Your input will be brought to the attention of the Technical Review Committee. Thank you for your assistance.

Instructors – If you found that additional materials were necessary in order to teach this module effectively, please let us know so that we may include them in the Equipment/Materials list in the Instructor's Guide.

Write: Curriculum Revision and Development Department
National Center for Construction Education and Research
P.O. Box 141104, Gainesville, FL 32614-1104

Fax: 352-334-0932

E-mail: curriculum@nccer.org

Craft ______________________ Module Name ______________________

Copyright Date ____________ Module Number ____________ Page Number(s) ____________

Description

____

____

____

(Optional) Correction

____

____

(Optional) Your Name and Address

____

____

Module 62206-02

Pipeline Maintenance

(CT 18 and 52)

Introduction to the Pipeline Operator Qualification Rule, API, and NCCER

It could be said that petroleum products keep America running. Petroleum products like oil and natural gas are the principal sources of energy in the United States and are used in many manufacturing and chemical processes. Petroleum products are primarily transported by pipeline. There are more than a million miles of liquid and natural gas pipeline in this country, ranging in diameter from two inches to four feet. These pipelines bring crude oil and natural gas from production fields to processing plants and refineries—and ultimately to the consumer.

Pipeline Operator Qualification

Petroleum products are considered hazardous, so pipeline personnel must take extra precautions and follow specific procedures when working with them. To help ensure the safe handling and transportation of petroleum products, the United States Department of Transportation (DOT) mandates safe practices as federal regulations.

One such regulation, the Operator Qualification (OQ) rule, requires pipeline operators to develop a written qualification plan to evaluate their personnel and their contract personnel (see *49 CFR Part 192 Subpart N—Gas,* and *49 CFR Part 195 Subpart G—Liquid*). The evaluation hinges on the operator's ability to perform certain covered tasks and to be able to recognize and respond to abnormal operating conditions that may be encountered while performing these covered tasks.

Covered Tasks

A pipeline performance task is deemed *covered* if the activity passes the following four-part test:

1. Is it performed on a pipeline facility?
2. Is it an operations or maintenance task?
3. Is it performed as a requirement of *49 CFR Part 195 (Liquid)* or *49 CFR Part 192 (Gas)*?
4. Does it affect the operation or integrity of the pipeline facility?

If the answer to all four questions is *yes*, then the activity is considered *covered* and must figure into the evaluation process of pipeline personnel.

American Petroleum Institute

The OQ rule and its related covered tasks provided the catalyst to develop a unified pipeline training and assessment program—so that pipeline operators and owners have ready access to the assessment and training tools they need. This effort was the direct result of the vision and leadership provided by The American Petroleum Institute (API) and API members.

Efforts to unify the industry are nothing new to API. Indeed, the impetus for forming the Institute in 1919 was the need to standardize engineering specifications for drilling and production equipment. Today, the most pressing issues revolve around public perceptions and government policies toward the industry. Speaking with one voice on these issues has become as essential as having interchangeable parts in the field. API's members determine the petroleum industry's positions on public policy as well as the standards that govern its day-to-day operations worldwide.

Consortium on Operator Qualification

API assembled a focus group whose primary purpose was to identify covered tasks and provide compliance recommendations to pipeline operators.

The Consortium on Operator Qualification (COOQ) became the guiding force for the identification and development of covered tasks for the liquid pipeline industry. Their work product, API publication 1161, *Guidance Document for the Qualification of Liquid Pipeline Personnel*, formed the basis for the resulting Pipeliner Training and Assessment Program.

The covered task list created in API publication 1161 has now evolved to include gas covered tasks and is maintained by NCCER. The most current covered task list can be downloaded at www.nccer.org.

National Center for Construction Education and Research

With the guidance and support of API, its members, and the Consortium on Operator Qualification, the National Center for Construction Education and Research (NCCER) was able to design and implement the Pipeliner Training and Assessment Program.

NCCER is a not-for-profit 501(c)(3) education foundation established in 1995 by the world's largest and most progressive construction and maintenance contractors and national contractor associations. NCCER was founded to address the

severe workforce shortage facing the construction and maintenance industries and to develop a standardized training process and curricula. Today, NCCER is supported by hundreds of leading construction and maintenance companies, manufacturers, and national associations and developed more than 35 construction and maintenance titles.

NCCER also maintains a National Registry, which contains the records of trainees and instructors who have received NCCER training. All completion information for every trainee is sent to the NCCER and filed within the Registry. The Registry can then confirm training and skills for workers as they move from company to company, state to state, or even within their own company.

Standardized Curriculum (NCCER's Contren™ Learning Series)

NCCER curricula are developed using the expertise of industry subject matter experts (SMEs). The use of SMEs as an integral part of every new development or revision ensures that the curriculum is current, industry-driven, and indicative of industry practices. The development and/or revision process is facilitated from start to finish by NCCER project managers, who work closely with the SMEs and a select team of technical writer(s) on the development and presentation of the curriculum.

Pipeline: Five Training Paths

The pipeline curriculum focuses on essential skills and competencies within five skill areas:

- Corrosion Control
- Electrical and Instrumentation
- Maintenance
- Mechanical
- Operations (Field and Control Center)

SMEs from the pipeline industry have determined the skills and competencies needed based on API's covered tasks and the consensus of the group/industry. Specific company information is generally not incorporated into the curriculum unless the SMEs determine that the information is the industry standard.

Training Modules and Covered Tasks

The smallest unit of instruction in the curriculum is a *module*. Each curriculum is composed of modules (the equivalent of stand-alone chapters) that are logically sequenced into complete *levels* of instruction. Every module is task-oriented and requires the successful completion of a written (knowledge-based) examination and a performance (hands-on) test. Modules may be individually selected and/or combined for efficient task training. Modules in the pipeline curricula, to the extent possible, will align with API-developed covered tasks. While not all modules will relate to covered tasks, all covered tasks will relate to at least one module.

Benefits of Training to the Pipeline Marketplace

Applying NCCER's standardized training process allows pipeline owners and contractors in the United States to establish the benchmark for acceptable skill and competency levels within the pipeline industry. This process will assist training organizations in developing and/or improving standardized pipeline training programs that effectively match their resources and operating conditions while responding to regulatory demands and to the needs of the industry. The process will also recognize training organizations that meet the minimum standards of competence, warranting regulatory, public, and professional confidence. This in turn will foster national unity within the pipeline industry for the mutual benefit of individual technicians, owners, contractors, and ultimately pipeline-users and society.

Additional Resources

To review the Pipeliner Training and Assessment Program Implementation and Resource Guide and the latest version of pipeline covered tasks, visit NCCER's Web site at www.nccer.org, or contact:

National Center for Construction Education and Research
PO Box 141104
Gainesville, FL 32614-1104
(888) 622-3720
Web site: www.nccer.org

For more information on API, please contact:

American Petroleum Institute
1220 L Street, Northwest
Washington, D.C. 20005
(202) 682-8000
Web site: www.api.org

To order *API Pub 1161*, contact Global Engineering Documents by calling 1-800-854-7149 or visiting their Web site at www.global.his.com.

Additional sources of information on the pipeline operator qualification regulation include:

DOT – Office of Pipeline Safety
www.ops.dot.gov

Transportation Safety Institute
www.tsi.dot.gov

COURSE MAP

This course map shows all of the modules in the second level of the Pipeline Maintenance curriculum. The suggested training order begins at the bottom and proceeds up. Skill levels increase as you advance on the course map. The local Training Program Sponsor may adjust the training order.

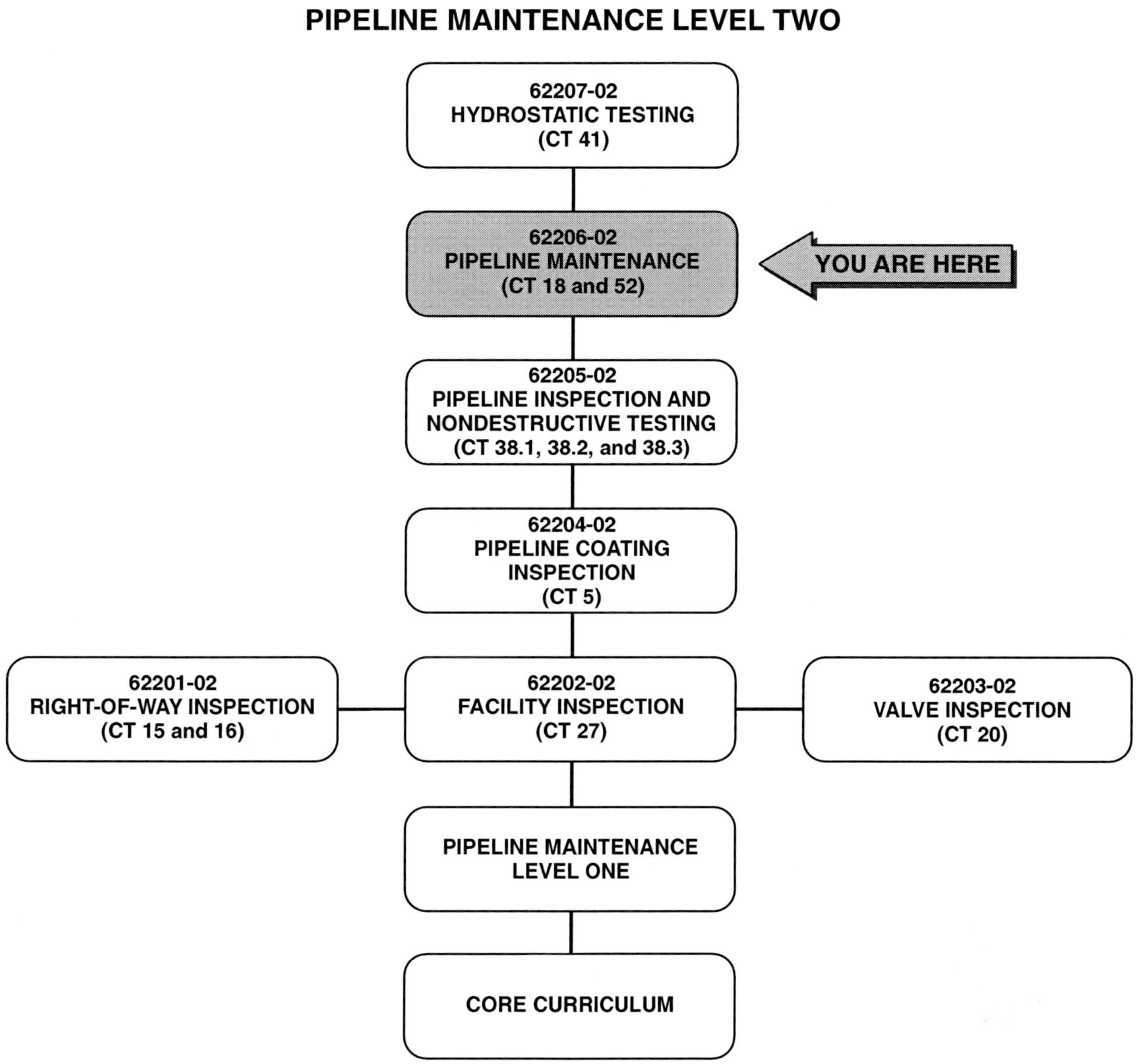

206CMAP.EPS

MODULE 62206-02 CONTENTS

Figures

Table

MODULE 62206-02

Pipeline Maintenance (CT 18 and 52)

Objectives

When you have completed this module, you will be able to do the following:

1. Locate underground structures
2. Maintain the pipeline route.
3. Excavate utilities and pipeline components.
4. Apply coatings.
5. Perform flange tightening.
6. Run pigs and perform maintenance pigging.
7. Inspect the pipeline following excavation activities, and perform a leakage survey after blasting (CT 18).
8. Conduct a leakage survey of a natural gas pipeline (CT 52).

Prerequisites

Before you begin this module, it is recommended that you successfully complete the following: Core Curriculum; Pipeline Core, Modules 66101-02 and 66102-02; Pipeline Maintenance Level One; Pipeline Maintenance Level Two, Modules 62201-02 through 62205-02.

Required Trainee Materials

1. Pencil and paper
2. Appropriate personal protective equipment

1.0.0 ◆ INTRODUCTION

In the past, pipeline maintenance usually meant pipeline repair. Repair crews were kept at various locations throughout the pipeline system in order to respond to emergency situations, usually when leaks were discovered.

Leaks were commonly caused by failure in joints, such as collars; pipe splitting when lap welds fail; corrosion pit holes; pipe ruptures; and third-party damage. Additionally, stresses on the line were caused by the pressure pulsations of reciprocating pumps. Repair often required shutting down the line to replace damaged pipe or make permanent repairs, as well as to clean up and dispose of spilled product.

Current pipeline maintenance procedures focus on the prevention of damage and deterioration so that emergency leaks and repairs can be reduced. Damage and deterioration were often considered normal wear and tear in the past, but with improvements in materials, construction, equipment, and operating and maintenance procedures, this has changed. Many of the causes of pipe failure have been eliminated or greatly reduced by smart pigging, One-Call systems, and the use of better materials.

Pipeline maintenance procedures are covered under Department of Transportation (DOT) regulations and have been carefully designed to prevent pipeline damage and failure. This module discusses the following elements of pipeline maintenance:

- Locating underground structures
- Maintaining the pipeline route
- Excavating utilities and pipeline components
- Applying coatings
- Performing flange tightening
- Running pigs
- Performing maintenance pigging

2.0.0 ◆ LOCATING UNDERGROUND STRUCTURES

Precise location of underground structures before any excavation begins is a critical part of maintaining a pipeline. Not only will digging in the wrong place be a waste of time and expense, but it may also inadvertently damage the pipe and/or other structures. Location of the company's pipeline should be clearly shown on company alignment maps, which are to be used along with visible markers.

However, the company may not be the only pipeline or utility with structures in the area, so any plan to excavate must be preceded with alerting the One-Call system. When this is done, any company with buried lines in the area of planned excavation will be alerted and asked to provide temporary markings and information, such as depths and line descriptions. The One-Call number is shown on permanent marker signs in the area.

2.1.0 Resources

Several resources are available to assist in locating underground structures. These are:

- One-Call system
- Maps and drawings
- Markers
- **Pipe locators** and cable locators

2.1.1 One-Call System

Excavations require notifying local utilities and having the underground line locations identified. However, the most common method used is an operator-sponsored centralized program known as the One-Call system. The One-Call system is defined as a communication system established by two or more utilities, governmental agencies, or other operators of underground facilities to provide one telephone number for excavating contractors and the general public to call for notification of their intent to use equipment for excavating, demolition, or other work. It provides the participating members an opportunity to identify and locate their facilities, to approve plans for others excavating, and to be present during excavations.

Notification and utility location can be handled either by utility operators individually or by a collaborative multicompany effort. Some operators may elect to set up their own system for handling calls from excavators and the location and marking of underground lines.

2.1.2 Maps and Drawings

Maps supplied by the pipeline company should indicate the route of the pipeline or structure with relation to surface features. Some will also show the depths. These maps can be of several different types or combinations of types. As-built survey drawings should show the route and the profile of the pipeline with the original depth recorded.

2.1.3 Markers

Permanent markers are required to be present at intervals along any pipeline or other structure that may be damaged by excavation. Operators are also required to provide temporary markers of all structures in the areas of excavation activity. These should remain in place as long as the excavation activities are continuing.

2.1.4 Pipe and Cable Locators

Equipment for locating pipe includes probe bars, or **probe rods**, and electronic or ultrasonic devices. A probe bar is a steel or fiberglass rod that generally has a T-handle attached. The rod is pushed into the ground until contact is made with the pipe. The probe bar is used to determine the location and depth of pipelines. Electronic and ultrasonic locators can also be used to locate the route of the pipe/structure, and some of them will indicate depth as well.

2.2.0 Procedures

Follow these steps to locate an underground structure:

Step 1 Identify the section/structure to be located.

Step 2 Locate the most recent version of the maps and alignment drawings to check the route location.

Step 3 Review the as-built survey information for depth information.

Step 4 Determine access requirements because some rights-of-way (ROWs) are restricted.

Step 5 Check the operation of locating equipment. Calibrate it if necessary.

Step 6 Use equipment to locate the route of the line and/or structure.

Step 7 Use the equipment, drawings, or probe rods to determine the depth of the structure.

> **NOTE**
> Probing is done across the width of the pipe at intervals less than one-half the pipe diameter. When the pipeline is located, probes are stuck in the ground along its length as a guide for excavators. If the pipe is too deep to reach with the probe, machine excavation is performed at one-half the probe depth. At that point, the line is re-probed until located and confirmed.

Step 8 Document the depth findings.

Step 9 Provide temporary marking for excavation crew.

> **NOTE**
> Temporary markers typically include paint on the ground and/or flags.

3.0.0 ◆ MAINTAINING THE PIPELINE ROUTE

Maintaining the pipeline route is necessary in order to access the pipeline for inspection, maintenance, and repairs. Where the pipeline and its equipment are aboveground, they must also be maintained. Markers on the route are required by regulation to inform the public and possible excavators of the existence of the pipeline. They are also used by aerial and other patrols in locating the pipeline accurately.

Pipeline route maintenance includes:

- Maintaining overhead crossings
- Installing permanent markers
- Maintaining the right-of-way (ROW)

3.1.0 Maintaining Overhead Crossings

When underground crossings are not practical, overhead crossings are constructed. Problems involving the strength and flexibility of the pipe and supporting structures not ordinarily encountered in pipeline construction must be considered. Strength requirements vary because of wind and ice loads. Flexibility must be provided to accommodate dimensional changes in the pipe and structures from temperature and load variations. Vibrations that cannot completely be eliminated by structures provide other stresses on the pipeline. Maintenance of these crossings requires monitoring and reducing the effects of these stresses. A major part of this is the installation, replacement, and repair of support structures.

Follow these steps to install, replace, or repair support structures:

Step 1 Determine the proper construction, spacing, and design of the pipe support, including the supporting structure foundation, company practice, and engineering design standards.

Step 2 Obtain the proper tools and equipment to complete the installation.

Step 3 Install a temporary support if applicable.

Step 4 Lower the internal pressure to 50% of maximum operating pressure (MOP) if pipe movement is required.

Step 5 Lift the pipe using the proper techniques.

Step 6 Remove the existing pipe support if applicable.

Step 7 Inspect the pipe wall.

Step 8 Repair the pipe support to meet standards if it is to be repaired.

Step 9 Install the repaired, new, or replacement support and insulator.

Step 10 Lower the pipe to the desired position.

Step 11 Remove the temporary pipe support.

Step 12 Adjust the support height to ensure a level pipe if an adjustable support is being used.

3.2.0 Installing Permanent Markers

Markers of various kinds have been used to show locations of pipelines at such places as road, railway, and waterway crossings. DOT regulations require that specific information be shown on these permanent markers and that additional markers be installed at other points.

These markers are often damaged by the elements or by the activities of others and must be repaired or replaced. Markers and signs to guide aerial patrols must also be installed and maintained along routes patrolled by aircraft. *Figure 1* shows examples of cross-country ROW marker positioning. *Figure 2* shows cross-country ROW markers and signs.

The installation of permanent markers covers:

- Types of posts
- Buried line markers
- Aboveground facility markers

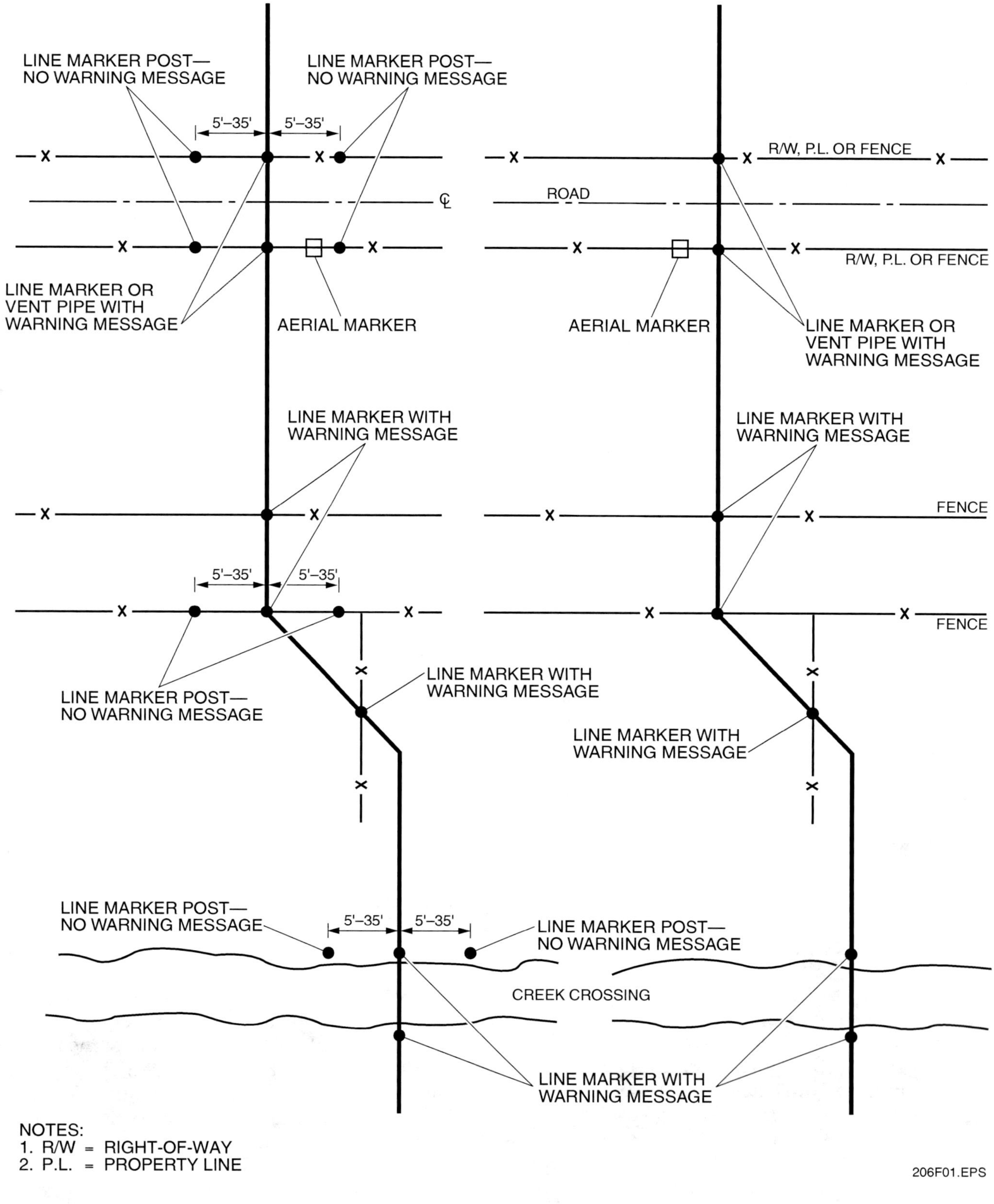

Figure 1 ◆ Cross-country ROW marker positioning.

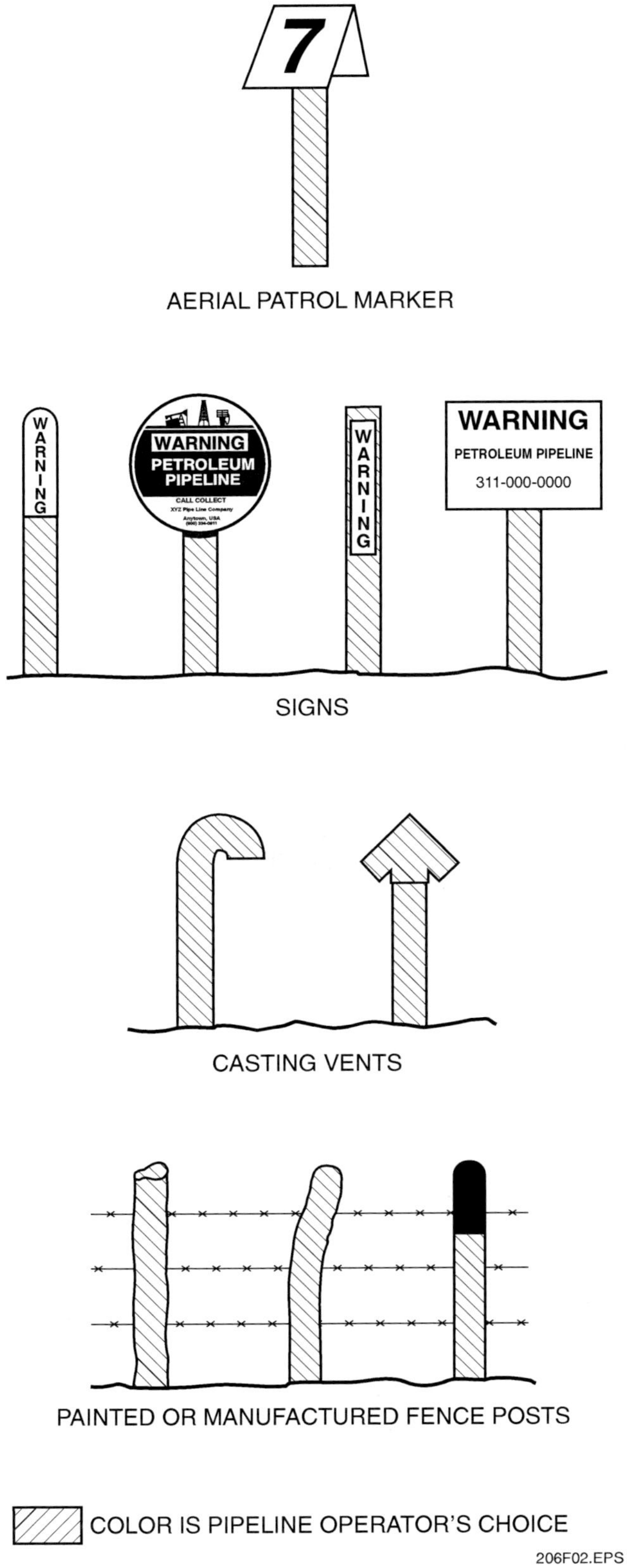

Figure 2 ◆ Cross-country ROW markers and signs.

3.2.1 Types of Posts

Posts may be of any material that will ensure adequate strength, stiffness, visibility, and durability. To maintain structural integrity and appearance, some post materials require surface protection against above- and belowground corrosion or weathering. A proven coating system that provides a suitable finish and nonfading color should be selected for this purpose. The following criteria should be applied in the selection of marker posts:

- Metal pipe posts should be straight, sound, and have a nominal diameter of 2 inches or more.
- Metal structures designed for use as posts may be used.
- Straight posts made from debarked trees and treated with a pressure-applied chemical preservative may be used. The smaller end of such posts should not be less than 8 inches in diameter. Wood posts are not recommended where brush or grass fires may be expected.
- Square precast reinforced concrete posts having a minimum cross-sectional area of 16 square inches may be used. Special conditions, such as spalling during freeze and thaw cycles, should be considered when specifying the material for these types of posts.
- Posts made of PVC, polyethylene, and fiberglass may be used. Material should, however, be resistant to ultraviolet exposure and suited to the environment where they will be used.
- Other materials are acceptable provided they meet the general criteria above. For example, **vent** pipes, fences, fence posts, or other existing posts may be used if they meet the criteria.

3.2.2 Buried-Line Markers

The information displayed on line markers must be presented in lettering of a certain size and stroke. The message should be presented on strong, durable material finished to resist the effects of exposure and vandalism. The minimum message is *WARNING, PETROLEUM PIPELINE.* The lettering must be at least 1 inch high on a background of sharply contrasting color. It must also contain the name of the pipeline operator with a telephone number, including an area code, where the operator can be reached at all times. The dimensions, wording, colors, and configuration must provide good visibility. *Figure 3* shows appropriate line markers.

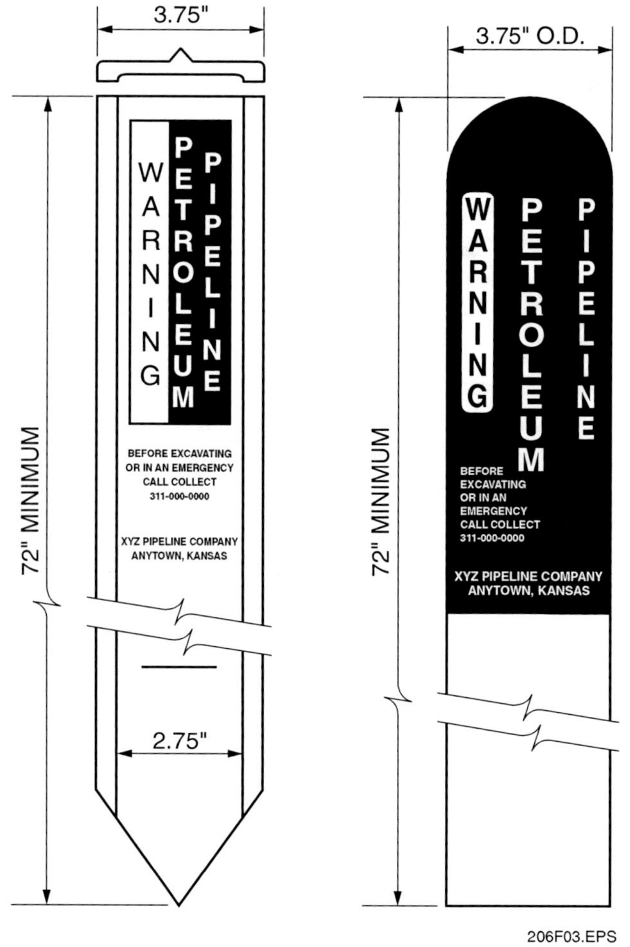

Figure 3 ◆ Appropriate line markers.

Other markers used for buried lines may include:

- Aerial patrol markers
- Prominently colored posts at fences and ROW limits of roads and railroads
- Markers at banks of water crossings
- Stenciled markings on the surfaces of pavements
- Buried tape
- Casing vent signs and marking
- Cathodic protection station signs

When signs are used to identify pipeline crossings at navigable waterways, they must contain the words, *DO NOT ANCHOR OR DREDGE*. The sign's lettering must not be less than 12 inches high. *Figure 4* shows a waterway crossing sign.

3.2.3 Aboveground Facility Markers

Pipeline facilities, such as stations, terminals, tank farms, valves, and metering or pipeline junction manifolds, must be marked with appropriate signs. These signs must include appropriate information as discussed earlier and may contain other data. Pipeline facility signs should be placed on all sides of the facility, at facility entrances, and in sufficient numbers so that the facility is clearly identified. *Figure 5* shows examples of facility signs.

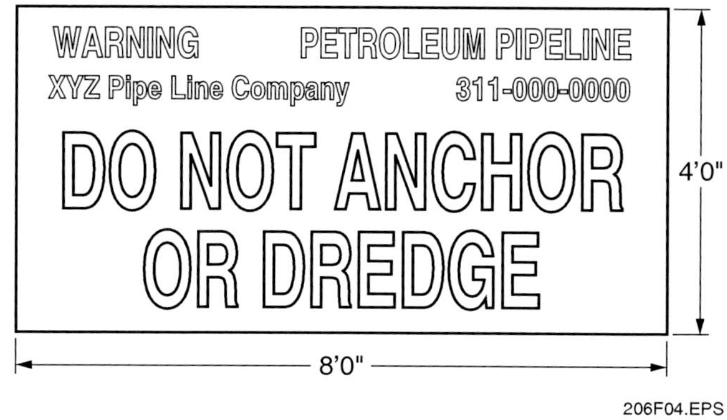

Figure 4 ◆ Waterway crossing sign.

3.3.0 Maintaining the Right-of-Way

Various maintenance work is required for the ROW. Protective earth cover must be maintained against earth movement and erosion. Markers must be kept in effective condition. The area over the pipeline must be free of vegetative growth that could interfere with the pipeline or access to the pipeline. Maintenance of the ROW includes:

- Maintaining earth cover
- Maintaining markers
- Clearing brush and growth

3.3.1 Maintaining Earth Cover

Maintaining earth cover is a major maintenance requirement in many areas. Soil erosion by water and wind and washouts by floodwaters may expose or threaten to expose the pipelines. Preventive measures include terracing to control surface erosion, building fences to control drifting sand, and building bulkheads to control river and stream bank erosion. Corrective measures, such as digging deeper ditches and lowering the pipelines, are expensive, and there is always the possibility of damaging the pipe coating or the pipe. The amount of earth cover should be checked to ensure that it meets the requirements of the company policy for the location.

3.3.2 Maintaining Markers

Periodic inspection of the ROW should indicate any problems with markers. Maintaining markers

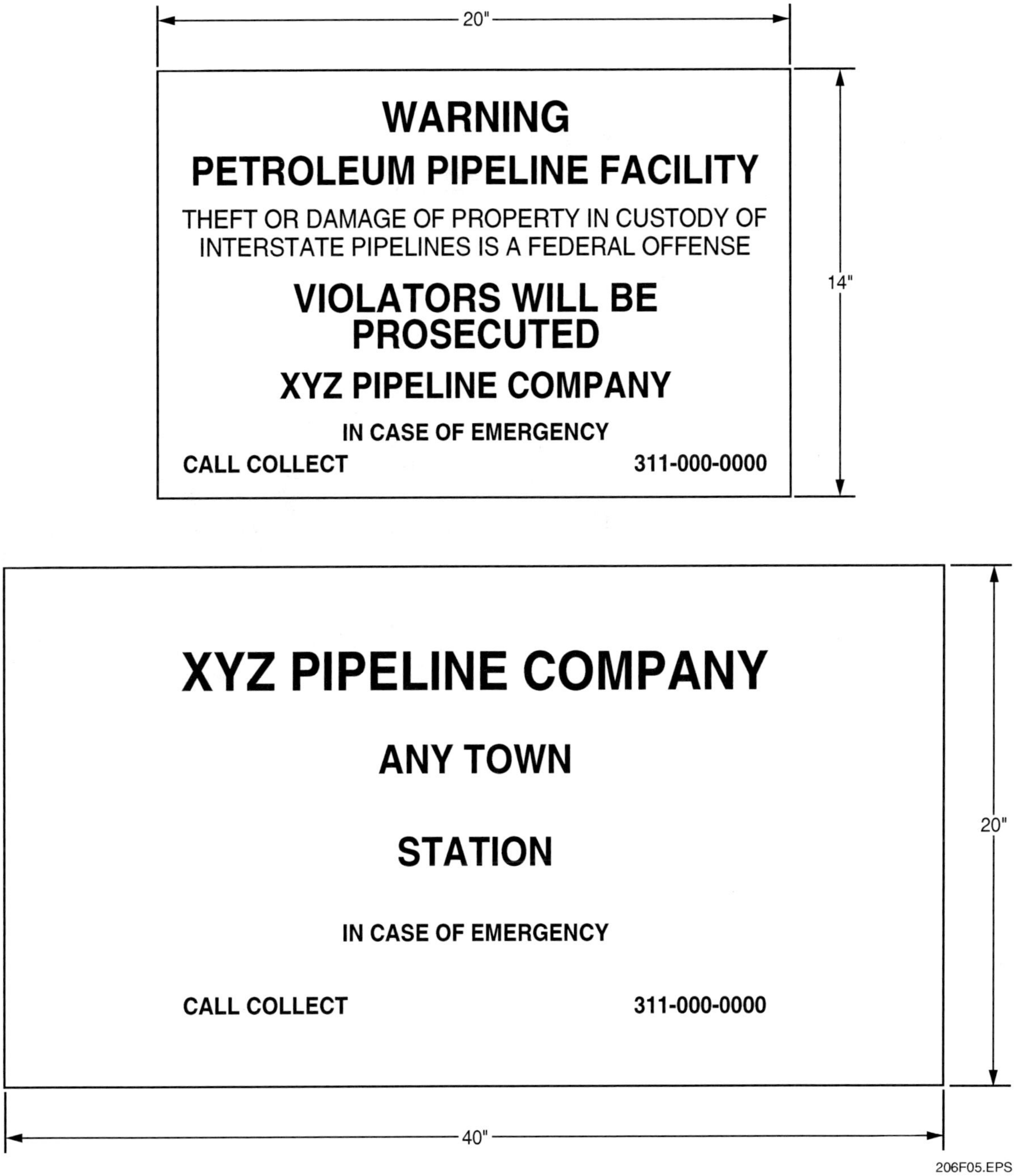

Figure 5 ◈ Examples of facility signs.

may include replacing, repairing, moving, or adjusting markers that are:

- Missing
- Damaged
- Showing incorrect information
- Misplaced
- Not properly visible
- Not properly oriented

Supports for markers may also need to be repaired, replaced, or repainted.

3.3.3 Clearing Brush and Growth

Vegetation can interfere with access to the pipeline as well as interfere with the pipeline itself. It can also cover up markers. All vegetation, such as bushes, trees, and undergrowth, should be cleared from the ROW over the pipeline before it can cause problems and before excessive digging becomes necessary. Herbicides may be used if they are appropriate for the location and if they meet local restrictions; their use must be restricted around waterways.

4.0.0 ◆ EXCAVATING UTILITIES AND PIPELINE COMPONENTS

Sometimes it becomes necessary to excavate utilities or pipeline components. Any time excavation is performed, procedures are followed to ensure that the pipeline is not damaged and that other underground structures and utilities are not disturbed or damaged. The site is usually stripped with backhoes and uncovered by hand. The pipe component, if removed, is taken up with the proper equipment.

Before beginning this section, review the Maintenance Level One module on excavating and backfilling for an introduction to OSHA requirements, hazards, One-Call systems, sloping, shoring, and other excavation information.

This section includes:

- Excavating utilities/pipeline components
- Backfilling

4.1.0 Excavating Utilities/Pipeline Components

Excavating utilities and/or pipeline components includes:

- Preparation
- Pipe location
- Excavation in Class B or C soil
- Excavation in Class A soil (rock)
- Excavation with a trench box (shield)

4.1.1 Preparing

The following is an example of the typical preparation for excavation. Always follow company policies and procedures. Follow these steps to prepare for excavation:

Step 1 Ensure that the One-Call system for the location has been contacted and that the ticket has been cleared and has valid digging dates.

Step 2 With maps and markings, locate the facility to be excavated and any utilities in the area.

Step 3 **Daylight** all utilities if possible, and verify their depth and location.

Step 4 Ensure that a pipeline company designated representative is present during all pipeline excavation activities.

4.1.2 Locating the Pipe

The following is an example of a typical pipe location procedure. Always follow company policies and procedures. There are many different methods for locating pipe. The following steps show one method used to locate and mark the pipeline:

Step 1 Use a pipe locator to find the pipe and to determine its approximate depth.

Step 2 Install two probe rods in two locations in line with the pipe, and verify the locator readings.

> **NOTE**
> Sometimes the pipe is deeper than the probe rods can reach without first removing some of the overburden.

> **CAUTION**
> Do not dig until you are certain of the pipe location, unless about 1 foot of cover can be safely removed at a time until locating the pipe.

Step 3 If the line is located with the probe rods, place them approximately 2 to 3 feet apart, parallel with the pipe, and to the side of the pipe nearest to the backhoe bucket.

Step 4 Place the excavator parallel with the pipe and off to one side, using the probe rods as a guide to ensure the excavator is straight with the pipe.

> **NOTE**
> This is to prevent the excavator from pulling the bucket into the pipe while digging. Another method is to place the excavator where the bucket is pulling slightly away from the pipe as it digs.

4.1.3 Excavating in Class B or C Soil

The following is an example of a typical excavation in Class B or C soil. Always follow company policies and procedures. Follow these steps to excavate the pipe:

Step 1 Start excavating the line, having someone probe the digging area after removing 1 foot of soil.

> **CAUTION**
> All excavated soil must be placed at least 2 feet away from the edge of the ditch.

Step 2 Continue excavating 1 foot at a time until you are within 2 feet of the top of the pipe.

Step 3 Have someone daylight the top and side of the pipe.

> **NOTE**
> Probe rods should remain in the ditch to indicate the side of the pipe.

Step 4 Excavate to within 1 foot of the top and side of the pipe after it has been daylighted.

> **NOTE**
> As the hole is dug, care must be taken to maintain the proper slope on the sides of the ditch as defined by OSHA standards. Someone must also be allowed access to the excavation to probe and daylight the pipe and must have a runout at the beginning of the ditch.

Step 5 Continue excavating, while keeping the pipe exposed, until the bottom of the ditch is 3 feet below the bottom of the pipe.

Step 6 Reposition the excavator as before on the other side of the pipe.

Step 7 Remove the soil on this side in the same manner to within 1 foot of the top and side of the pipe and to 3 feet below the bottom of the pipe.

Step 8 Have the digging crew manually remove the 1 foot of soil on either side of the pipe.

Step 9 With the excavator, remove the soil left by the digging crew to return to the 3 foot depth below the pipe.

Step 10 Manually remove approximately 2 feet of soil from below the pipe, leaving this soil in the bottom of the trench.

Step 11 Place skids or some type of approved support under the pipe every 20 feet or as specified by the inspector.

4.1.4 Excavating in Class A Soil (Rock)

The following is an example of a typical excavation in Class A soil. Always follow company policies and procedures. Follow these steps to excavate the pipeline:

Step 1 Follow the steps for excavation in Class B or C soil with the following additional requirements as needed.

Step 2 Provide shielding for the pipeline with oak planks, blasting mats, or oversized pipes or sleeves.

Step 3 Use an excavator-mounted rock hammer to remove rock from the sides of the pipe.

Step 4 Use a handheld jackhammer and a rock splitter to remove rock from under the pipe if necessary.

Step 5 Fracture rocks too large to fit in the excavator bucket prior to removal to avoid damaging the pipe.

4.1.5 Excavating with a Trench Box

The following is an example of a typical excavation with a trench box. Always follow company policies and procedures. Follow these steps to excavate the pipeline with a trench box:

Step 1 Follow the steps for excavation in Class A, B, and C soil with the following additional requirements as needed.

Step 2 Ensure that the ditch is properly sloped to expose the pipe.

> **NOTE**
> With large diameter pipe, there is not enough room to set the trench box and dig inside of it. In extreme cases, as excavation continues down, a trench box can be temporarily set in the hole to protect the workers daylighting the pipeline and can then be removed to continue excavation.

> **CAUTION**
> When the trench box is temporarily placed in the ditch, soil should be placed around the box to keep it from moving in case of a cave-in.

Step 3 Once the pipeline is exposed, the shield can be placed in the ditch and locked in by backfilling around the box, keeping the soil at least 18 inches below the top of the box.

Step 4 Use end plates, sheet piling, or other methods suitable to retain the soil if either end of the ditch cannot be properly sloped.

Step 5 Leave the shield in place as the ditch is backfilled around the pipe.

> **NOTE**
> This step protects the workers while they place soil under the pipe.

Step 6 Remove the shield once the pipeline has a 2-foot cover.

Step 7 Backfill normally.

> **NOTE**
> If a ditch longer than the shield is required, the box can be moved up the pipeline. Excavate in front of the shield using the above guidelines, and use another box for the new ditch until the other box can be moved forward. Backfill behind the box before it is moved forward.

4.2.0 Backfilling

After maintenance or repairs, the soil must be replaced or backfilled into the ditch over the pipe. Backfill material must be free of rocks, roots, and any other foreign material that could damage the pipe coating or cause uneven loading of the pipe. The inspection begun after repairs must be continued during the backfill operation. Pipe must be properly supported when any temporary supports are removed, and backfilling must provide adequate cover for the pipeline. Follow these steps to backfill over the pipeline:

Step 1 Place proper supports, such as sand bags, under the pipe.

Step 2 Remove any skids or other temporary supports.

Step 3 Check the backfill material for rocks, roots, or foreign material, or ensure the backfill material has been approved by the inspector.

Step 4 Begin adding backfill to the ditch and pushing soil under the pipe.

Step 5 Tamp or compact soil under and around the bottom of the pipe with a tamping tool.

Step 6 Use the excavator to backfill until the pipe is completely covered with 2 feet of soil, filling the ditch equally along both sides of the pipe.

Step 7 Compact the soil as it is added to the ditch.

Step 8 Use a dozer or loader, running only perpendicular to the pipe, to finish backfilling.

Step 9 Compact the soil over the pipe when the ditch is filled.

Step 10 Crown the backfill over the pipe to allow for settlement.

> **NOTE**
> Add extra cover for procedures at road crossings, ditches, and streams. Allow extra compaction at road crossings and other locations where heavy loads may cross the pipeline.

5.0.0 ◆ APPLYING COATING

Pipe that is to be field coated must be prepared for the application. First, it must be cleaned of grease or oil with the use of solvents. Solvents that leave an oily residue, such as kerosene, should not be used. Brushes and scrapers may be needed to remove the dirt and rust. After cleaning, the pipe is primed and then the protective coating is applied. Coating application includes:

- Types of coating
- Primer application
- Coating application
- Coating protection

5.1.0 Types of Coating

Different coating types are available to meet the specific demands of the pipeline. Some of the more commonly used coatings are enamels, epoxy resins, and tapes.

Coal-tar (bituminous enamel) is derived from pitches or petroleum asphalts. It is widely available, inexpensive, and easy to apply. For these reasons, coal-tar is widely used. Enamel coatings include asphalts, coal-tars, greases and waxes, mastics, and asphalt mastics. Asphalts occur naturally and as a petroleum distillation by-product.

Coal-tars are formed from coal distillations. Grease and wax are also petroleum based and contain corrosion inhibitors and fillers. All enamel coatings are unsuitable for use in hydrocarbon-contaminated soil. In very cold temperatures, enamels require special precautions to prevent cracking.

Mastics are mixed from asphalt, sand, and fillers and provide thicker, more durable coatings with greater adhesion. Asphalt mastics have been used for a long time. They are mixtures of sand, crushed limestone, and fiber, and contain an asphalt binder. This is the thickest coating type and is often used offshore.

Powdered epoxy resins form a virtual skin over a steel surface when applied to heated pipe. The fusion-bonding process requires specialized equipment and exacting preparation, so these coatings are usually applied at the mill.

One big advantage of the epoxy resin coating is its relative thinness. Inspection of the pipe is easier since defects are not hidden under thick layers of coating. Epoxy resins are also strong enough to resist handling damage, are durable and flexible, and can be applied to interior pipe surfaces. They can also withstand high operating temperatures.

Polyethylene tape is often used for field-applied coating. Other tapes used for coating include polyvinyl, coal-tar base, and butyl-mastic. An advantage of tape coating is that it does not raise any environmental or health objections.

5.2.0 Primer Application

Ensuring the pipe is properly prepared for coating application is extremely important. The pipe surface must be free of all moisture, dirt, scale, rust, and debris. If not, the coating will adhere improperly and will not be effective. In the mill or the field, brushes and buffers may be used.

After cleaning, a thin coating of primer is applied to prepare the surface. The coating and wrapping are done in accordance with manufacturer's specifications. In the field, special machines are used for cleaning, priming, coating, and wrapping directly over the ditch in which the pipe is being laid.

Again, the pipe must first be cleaned by power brushing or sand blasting. Then the primer can be applied. When the primer dries, the coating is applied. It is critical that the joint with adjacent coating be waterproof.

5.3.0 Coating Application

Specific coatings have specific application requirements and procedures. Enamels are melted in dope pots and kept at the right temperature for use in coating. The liquid coating is stirred constantly to evenly distribute heat and to keep the liquid uniform before it is fed through a hose to the coating machine, where it is applied in an even layer. Wrapping is applied over the coating to protect it while it hardens.

Fusion-bond epoxy coatings require the pipe to be heated and also require special equipment and exacting procedures to be used. Because of this, fusion-bond epoxy coatings are not normally applied in the field.

Taping machines apply tape in the field by moving along the line and wrapping tape around the pipe. Tape is supplied from rolls mounted on spindles on the machine, fed to the pipe under tension, and wrapped in overlapping layers.

Since the ends of pipe to be welded are left uncoated, they must be coated in the field after welding. The same applies to parts of the pipe from which coating has been removed for repairs. A variety of processes and materials may be used. If the rest of the pipe is coated with coal-tar, then coal-tar, glass, and felt are usually applied to the weld joints or repaired section. Heat-applied epoxy powder can be used with thin-film coatings. If the coatings are compatible, almost any type of coating can be used on field joints or repaired sections.

5.4.0 Coating Protection

Pipe wrapping is the material applied on top of the coating to protect it from damage during handling. Wrapping materials are usually polyethylene, polyolefin, felt, fiberglass, or kraft paper. Kraft paper also protects pipe from ultraviolet rays if it is exposed to the sunlight.

Other factors that pose a hazard to the coating include the moisture content of the soil, contamination of the soil, soils that vary greatly from wet to dry, and extreme temperature variations. These factors must be addressed when selecting the type of coating to use and the type of wrapping to cover it.

In some areas, it is difficult to remove all rocks from the environment. In these cases, rock shields can be used for protection. Some type of protective barrier may also be used to protect the coating from damage when existing pipe is excavated.

6.0.0 ◆ TIGHTENING FLANGES

Proper stud tension, correct torquing pattern, and clean, rust-free studs are necessary for properly tightened bolted connections. Correctly tightened flanges are necessary to provide an evenly distributed clamping force around bolted connections to prevent leakage. This section addresses the following:

- Tools
- Preparation
- Procedure

6.1.0 Tools

A hydraulic bolt tensioning system provides the most accurate method of applying proper bolt stress. A hydraulic bolt tensioning system is strongly recommended for tightening pump flanges and other critical flanged connections. This system must be used according to the manufacturer's operating procedures.

An impact wrench will generally develop sufficient torque for studs 1.25 inches in diameter and smaller. A torque wrench adequate to measure torque to within 5% is required to ensure adequate bolt tension.

6.2.0 Preparation

The following is an example of a typical preparation procedure. Always follow company policies and procedures. Follow these steps to prepare for flange tightening:

Step 1 Ensure that the piping is free of foreign materials and construction debris.

Step 2 Examine the studs for correct size and material specification.

Step 3 Examine the studs for mechanical damage, such as burrs or nicks, dirt, or rust. Replace if necessary.

Step 4 Check that the studs have sufficient length for the tensioning tool.

Step 5 Ensure that the gaskets are the correct size, type, and material for the application.

> **NOTE**
> Used studs that are in good condition may be re-used. Never install a used gasket.

Step 6 Clean the flange faces using a steel scraper or wire brush.

Step 7 Wipe the flange faces with a clean rag, and check for surface damage, trueness, and out-of-parallel alignment.

Step 8 If appropriate, lubricate the following with a recognized low-friction lubricant, such as molybdenum-disulfide-based Moly-Paste C-670:

- Threads of studs
- Faces of nuts
- Contact area of flanges

6.3.0 Flange-Tightening Procedure

The following is an example of a typical flange-tightening procedure. Always follow company policies and procedures. Follow these steps to tighten flanges:

Step 1 Align the flanges using at least two line-up pins sized 2 mm smaller than the diameter of the bolt holes.

Step 2 Insert a new gasket, and install the studs and nuts in the remaining holes.

> **NOTE**
> If the pipe run is horizontal, install the bottom studs first to retain the gasket.

Step 3 Tighten all nuts handtight, taking care that the bolts or studs extend completely through the nuts.

Step 4 Tighten four studs at 90 degrees to each other until there is no visible gap at the gasket.

> **NOTE**
> It may be necessary to tighten more than four studs to close the gap.

Step 5 Measure the distance from face to face at the flange outside diameter, moving around the flange circumference to ensure that they come together evenly.

> **NOTE**
> The flange is now faced.

Step 6 Number the studs according to the numbering and torque diagram. *Figure 6* shows the numbering and torque diagram.

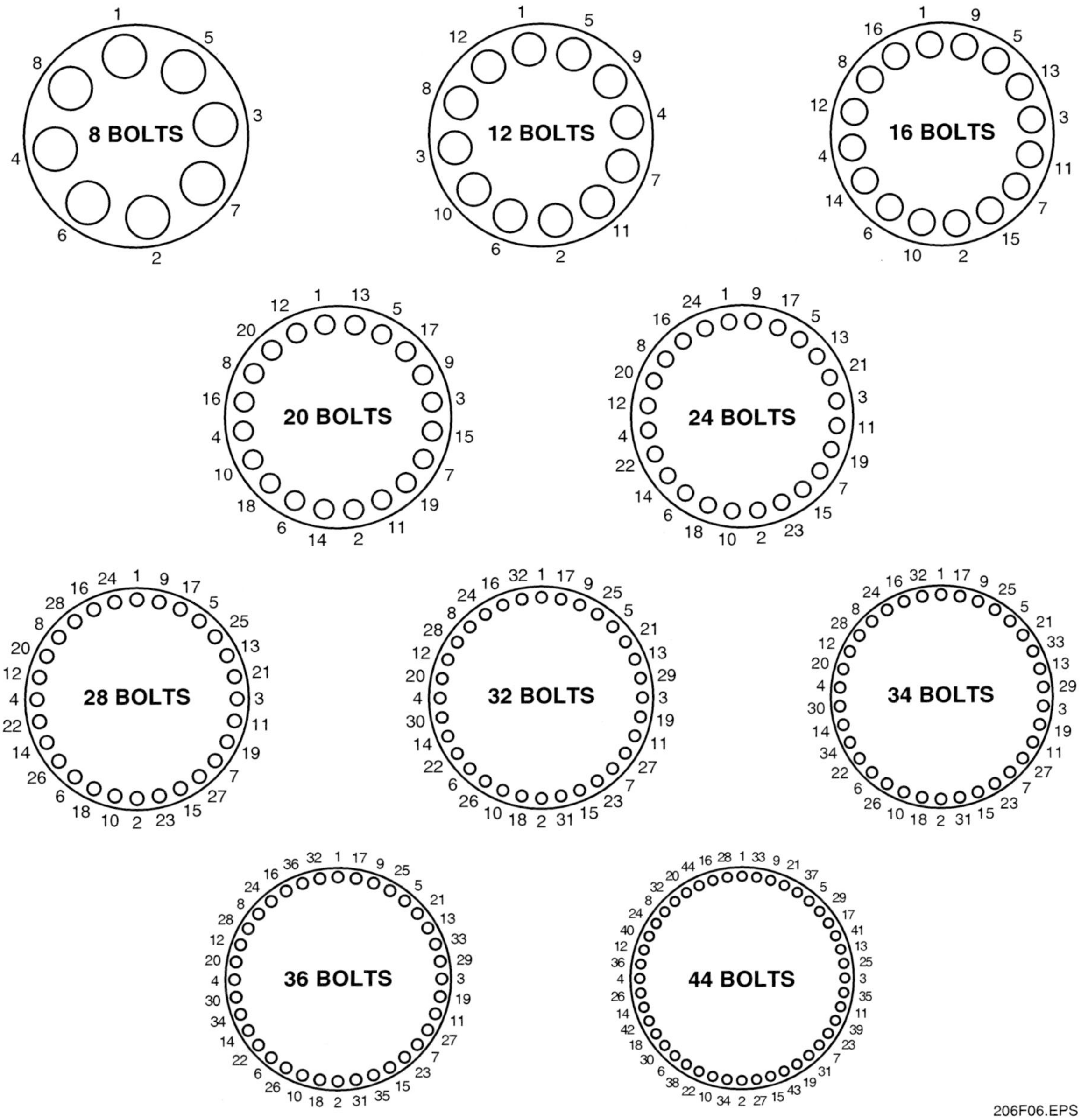

Figure 6 ◆ Numbering and torque diagram.

Step 7 Tighten the studs in numbered sequence (diametrically opposed) to approximately 30% of the torque value for the stud size. Skip the line-up pins; leave the pins in place to ensure that the flange is aligned.

> **NOTE**
> If tightening pump flanges, some variation in sequence may be necessary to avoid excessive pump shaft deflection. *Table 1* shows the appropriate torque values.

Step 8 Remove the line-up pins, and install the studs and nuts.

Step 9 Tighten all the studs in numbered sequence to approximately 60% of the torque value shown in the table.

Step 10 Tighten the studs in numbered sequence to 100% of the torque value.

Step 11 Check the torque of each nut, moving to each adjacent nut around the flange in clockwise rotational sequence. Use one or more passes to ensure that the studs are evenly tightened.

Table 1 Appropriate Torque Values

Stud Size (in.)		Torque (ft-lb)
0.750	$\frac{3}{4}$	175
0.875	$\frac{7}{8}$	300
1.000	1	450
1.125	$1\frac{1}{8}$	625
1.250	$1\frac{1}{4}$	875
1.375	$1\frac{3}{8}$	1,200
1.500	$1\frac{1}{2}$	1,400
1.625	$1\frac{5}{8}$	1,925
1.750	$1\frac{3}{4}$	2,625
1.875	$1\frac{7}{8}$	3,500
2.000	2	3,850
2.250	$2\frac{1}{4}$	5,600
2.500	$2\frac{1}{2}$	7,700
2.750	$2\frac{3}{4}$	9,275
3.000	3	12,225
3.250	$3\frac{1}{4}$	15,650
3.500	$3\frac{1}{2}$	19,250

206T01.EPS

Step 12 Slowly pressurize the facility to working pressure, and visually check for leaks.

Step 13 If there is a leak, check the stud tension. If the leak cannot be stopped by further tightening:

- Relieve the pressure
- Remove the studs and gaskets
- Check for flange face defects and gasket defects
- Make the necessary repairs or replacements
- Repeat the installation

> **NOTE**
> If the flange assembly is in a high-pressure or high-temperature application, retorquing may be necessary after startup to compensate for any relaxation or creep in the assembly.

7.0.0 ◆ RUNNING PIGS

Inside pipe surfaces require periodic cleaning to remove deposits of wax, rust, scale, or other materials. Removal may be to restore throughput capacity, or it may be preventive maintenance to control corrosion. A major component of inspecting and maintaining the pipeline interior is the pig. The following sections discuss pigs and pig traps.

7.1.0 Pigs

A variety of scraping devices called pigs, scrapers, go-devils, or displacers are used to clean pipeline interiors and to perform other inside functions. Smart pigs are instrumented to survey the interior and provide information on the interior surface condition. Smart pigs are available to perform several different tasks. Displacer pigs clean water or product out of a section of pipeline when necessary. Pigs are also used as batch separators.

Pigs come in different shapes and sizes depending on their function and the pipe size. Traditional scrapers used since the earliest days of pipelining are still preferred by many operators. *Figure 7* shows a traditional scraper.

The small notched wheels on the scraper are held by spring pressure against the inside of the pipe and serve to keep the scraper centered inside the pipe. They reduce uneven wear on the sliding elements of the scraper and increase the scraper

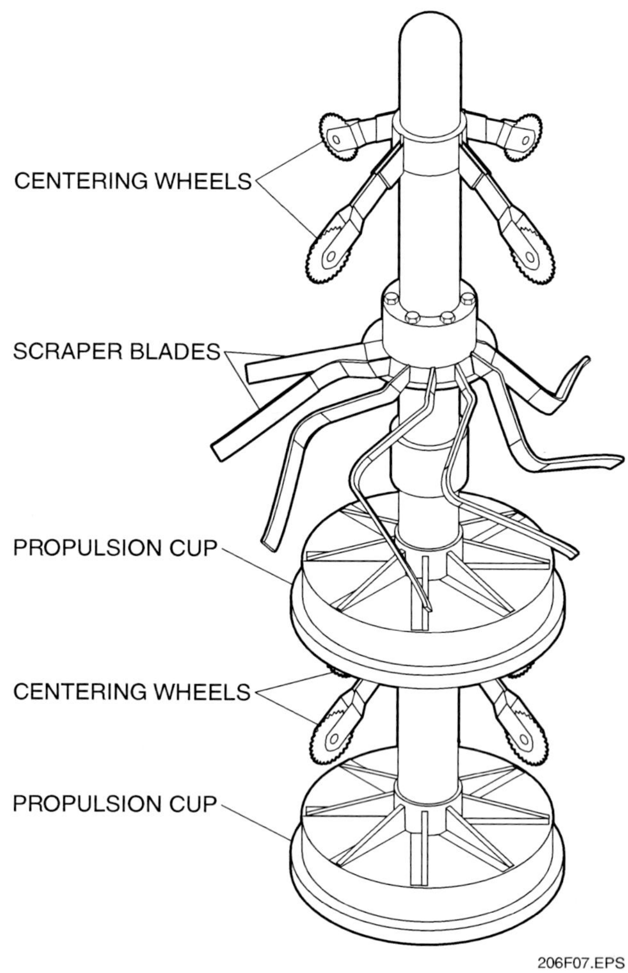

Figure 7 ◆ Traditional scraper.

noise, which helps to indicate the scraper's location. Curved blades scrape waxy deposits from the pipe. Other scraper types may use brushes for wax removal, but the brushes can become clogged with wax. Scrapers are usually fitted with a universal joint to allow it to go through pipe bends.

Scrapers used for large-diameter pipe can be quite heavy and use various methods to keep the scraper from wearing on the bottom side, including using a bypass stream of oil through the scraper to make the scraper rotate. The bypass also allows for dispersion of solids to prevent them from being compacted into plugs that pumps cannot move. Some large scrapers have a central tube filled with air to add buoyancy and to prevent uneven wearing.

Spherical scrapers made of oil-resisting plastic are generally used as batch separators rather than for maintenance, but they can sweep water and solids along. Some scrapers are designed specifically for use in pipe of different diameters.

7.2.0 Pig Traps

Pig traps are used for the insertion and removal of pigs in the pipeline. Valves in the system are operated by pig detectors, and open and close as needed to isolate the pigs from the main line or to launch them. *Figure 8* shows a pig trap system.

The increasing size of pipe and the increasing size of pigs has made it necessary to provide mechanical means of handling the pigs. Racks or ramps may be constructed at truck bed height at both the launching and receiving traps. Small cranes with winches may also be used.

Automation has extended to the launching of scrapers. Some traps are arranged so that one or more scrapers may be loaded into traps and then launched at different times by remote-controlled valves. At unattended locations, an arriving pig may signal the appropriate time for launching or initiate a programmed launching operation.

8.0.0 ◈ MAINTENANCE PIGGING

Maintenance pigging of pipelines is a major part of the maintenance work. Pigs may be run to clean out the line or to detect anomalies in the pipe wall. An anomaly is anything found by a smart pig in the pipe wall that causes a deviation of its normal signal. An anomaly could be a corrosion pit, a gouge, a dent, a crack, a thinning of the wall, a lamination, a blister, or any number of mill defects. These defects may or may not be injurious to the pipe and its integrity. An anomaly could also be a tap, fitting, or other attachment on the pipe.

This section covers the preparation and running of pigs. The success of a smart pigging run depends largely on the pre-run preparation. The job must be carefully planned.

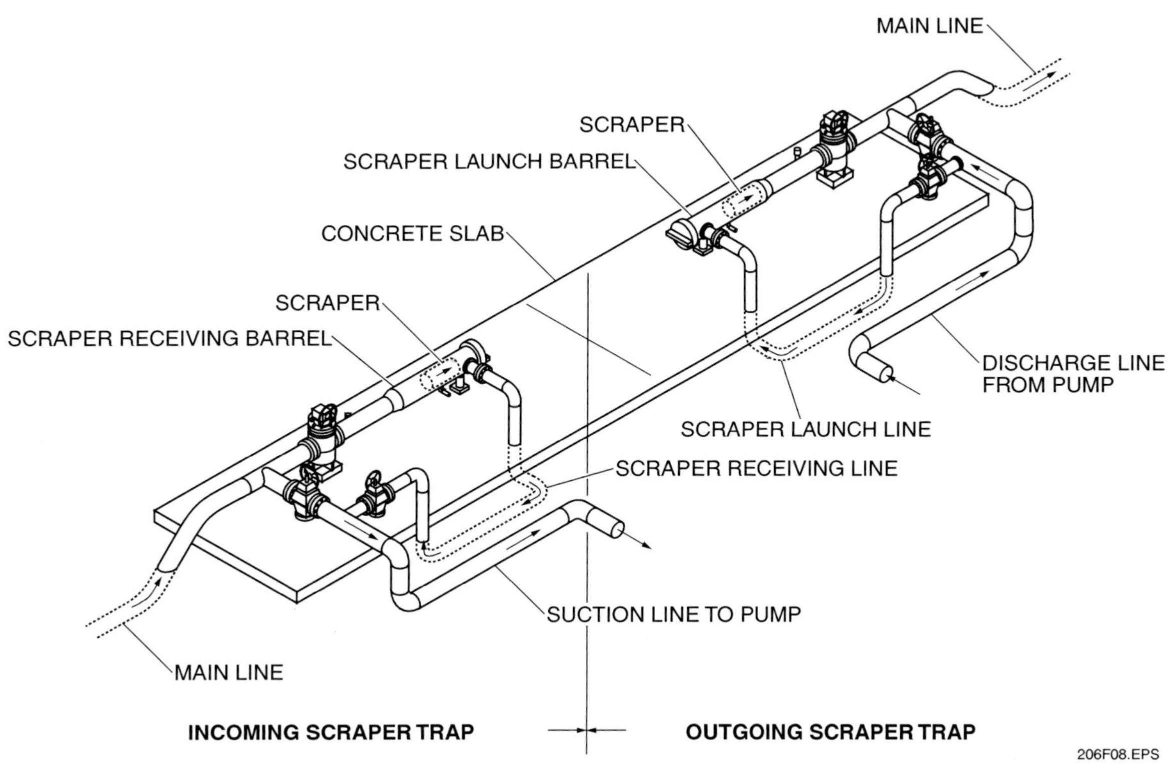

Figure 8 ◈ Pig trap system.

8.1.0 Preparing the Pipeline

The following is an example of a typical procedure. Always follow company policies and procedures. Follow these steps to ensure that the line is ready to accept the smart pig:

Step 1 Ensure that no restrictions exist in the pipeline that would cause the pig to become stuck:

- Match bend radii to the pig's capability.

> **NOTE**
> Pigs are typically chosen to negotiate a 1.5D bend or a 3D bend, but this is dependent on wall thickness.

- Verify that no reduced port valves exist in the line.
- Verify that the pig can pass check valves in the line.
- Locate and repair any dents, gouges, or other physical restrictions in the pipe. This may be done by running caliper pigs, deformation pigs, or pigs with sizing plates.

Step 2 Ensure that the full opening tees have scraper bars installed.

Step 3 Ensure that the scraper or pig traps have adequate length to accommodate the pig.

Step 4 Ensure that the pig trap valves close tightly without harmful leakage.

Step 5 Ensure that the line is clean and free of debris, pipeline rouge, and paraffin.

Step 6 Ensure that all downstream line block valves are fully open.

Step 7 Ensure that all side valves are closed if possible.

Step 8 Ensure that proper lifting equipment is available at the launch and reception of the pig.

> **NOTE**
> The lifting equipment may vary with the pig size and type.

Step 9 Predetermine the aboveground marker (AGM) locations, and mark the locations to withstand the elements and livestock where needed.

> **NOTE**
> AGM locations are typically no more than one mile apart.

Step 10 Ensure that the maps are prepared with the proposed AGM locations and are available to personnel.

Step 11 Ensure that adequate personnel are available to leapfrog the AGMs forward as the pig passes them.

Step 12 Ensure that adequate personnel and equipment are available to perform the company's role in launching and receiving the pig.

Step 13 Ensure that the line is clean by running a combination cup and disk pig one or more times. *Figure 9* shows a cup and disk pig.

> **NOTE**
> The pig should be in good condition and fit tightly in the pipeline when launched. Clean is defined as no more than one pound of waste for each 4 inches of nominal pipe diameter. One pint equals approximately one pound.

Figure 9 ◆ Cup and disk pig.

Step 14 If the line needs cleaning:

- Run less aggressive cleaning pigs, such as foam and/or cup pigs, if the line has not been cleaned in a while. *Figure 10* shows foam pigs.
- Run scraper and brush pigs for paraffin and other buildup. *Figure 11* shows a scraper and brush pig.

Step 15 Ensure that at least 1.5 times the calculated amount of oil is available to run the pig.

Step 16 Ensure that the pumps are in good condition and can provide steady flow at sufficient quantities to move the pig above its minimum travel speed.

Step 17 Ensure that the AGM crew understands their job and has means to communicate with each other as well as with the project manager.

Step 18 Notify the control center of the pigging plans and actual launch time.

8.2.0 Running the Tool

The actual running of the pig takes a joint effort by the pigging vendor and company personnel. Care must be taken to ensure that the pigging vendor's recommendations are followed regarding the handling of the pig to prevent damage. Follow these steps to run the pig:

Step 1 Ensure that all on-scene personnel are properly fitted with the correct personal protective equipment, including fire-retardant clothing (FRC) for those loading the pig into the trap and those removing the pig.

Step 2 Insert the pig into the trap, handling the pig according to vendor's instructions.

CAUTION

Forcing a pig into the trap can cause damage to the pig.

NOTE

Only company personnel operate company equipment associated with the launch.

Step 3 Listen for the pig traveling down the pipeline to ensure its launch.

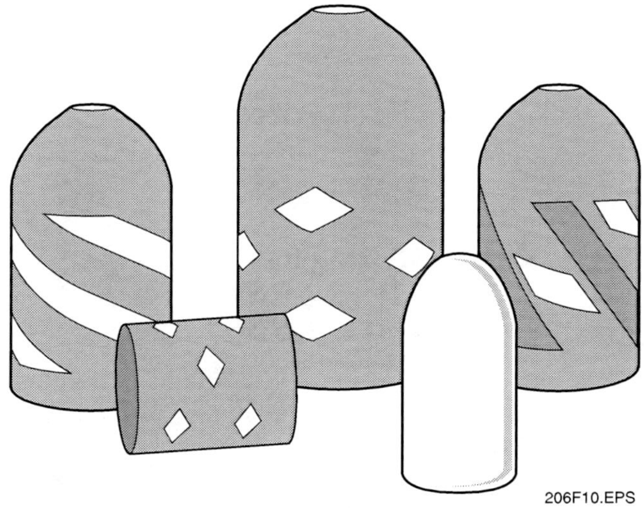

Figure 10 ◆ Foam pigs.

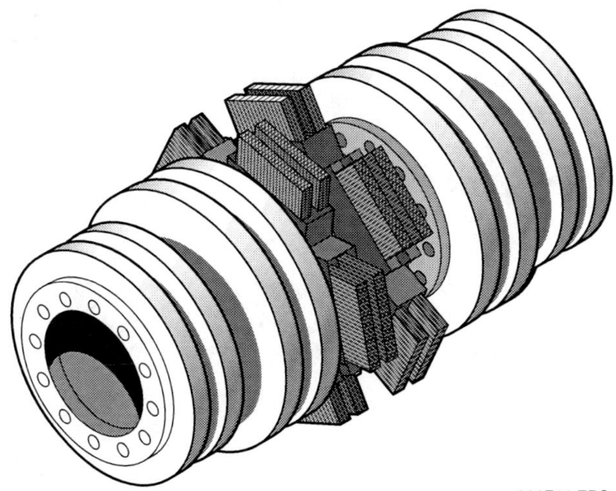

Figure 11 ◆ Scraper and brush pig.

Step 4 Post personnel at the first or next AGM well before the expected arrival time of the pig.

Step 5 Document the trigger time and location of each AGM.

Step 6 Compare the trigger times to the theoretical speed of the pig, based on flow rates.

Step 7 Trap the pig.

Step 8 Drain the trap.

Step 9 Remove the pig from the trap, following the vendor's instructions.

Step 10 Close the trap and resume normal operations.

After running the pig, the vendor furnishes preliminary results in a spreadsheet form or other medium. This may occur within a matter of hours or longer, depending on the vendor and the tool. Investigations of all anomalies are scheduled from this report.

9.0.0 ◆ PERFORMING POST-BLASTING LEAK SURVEILLANCE

After blasting operations have taken place near any section of pipeline, a leak survey must be performed. A leak survey is a visual inspection of the pipeline right-of-way directly affected by a blasting operation. An additional segment of pipe up- and downstream of the blast impact area must also be surveyed. The survey checks for evidence of leaks indicative of damage to the pipeline. Pipeline maintenance employees must be able to perform a leak survey in accordance with their company policy after any blasting has taken place in proximity of the pipeline.

> **WARNING!**
> While performing a leak survey, you may encounter an explosion or fire, such as burning hydrocarbon vapors. Protect yourself, the public, and other employees. Contact the local fire and police departments. Immediately initiate and execute line shutdown, and block, depressurize, and/or drain the affected line segment. Control and contain the leaking hydrocarbon liquids if you are qualified and trained to do so.

9.1.0 Performing a Leak Survey

A leak survey may include the use of **gas detection equipment**. It may also include exposure of the pipeline if, during the blasting operation, **seismographic vibration readings** are recorded in excess of the allowable limit as stated in the operator's specifications/policy.

> **CAUTION**
> Pipeline system damage, such as broken, bent, or burnt pipeline, excessive disturbance of earthen support, and dislodged pipeline, may be encountered during a leak survey. Immediately initiate and execute line shutdown, and block, depressurize, and/or drain the affected line segment. Perform site assessment if you are qualified and trained to do so.

Follow these steps to perform a leak survey:

Step 1 Identify the information or directing source determining the limits of the surface area to be visually inspected.

> **NOTE**
> It is necessary to ensure inspection of the entire area that may have experienced impact. Sources that might provide this information are the engineering department or an outside company blasting engineer.

Step 2 Determine to what the extent the inspection should be performed.

Step 3 Determine any equipment necessary to expedite the inspection. The terrain and environment along right-of-ways can vary greatly. Difficult or blocked access and egress to a site can interfere with proper inspection.

Step 4 Visually inspect the impact area for indication of a leak:

- Hydrocarbon liquid or frost on the ground
- Discolored or melted snow
- Hydrocarbon smell
- **Vapor clouds** or shimmery heat-wave-like product evaporation
- Blowing, whistling, or trickling sound

> **WARNING!**
> While performing a leak survey, you may discover an unexpected hazardous liquid or carbon dioxide. Crude, product, or a chemical may be blowing, leeching, boiling, or vaporizing into the ground, water, or air. Immediately take steps to initiate and execute line shutdown, and block, depressurize, and/or drain the affected line segment. Take any action to protect and minimize the danger to the public and environment, such as notifying residents and local police and fire departments to accomplish evacuation and to control traffic. Perform site assessment if you are qualified and trained to do so. Control and contain the leaking hydrocarbon liquids if you are qualified and trained to do so.

Step 5 Collect data about the existing ground surface condition.

Step 6 Determine if the post-blast surface condition indicates that exposure of the pipeline is necessary.

> **NOTE**
> Go to Step 7 if exposure is not necessary. Go to Step 8 if exposure is necessary.

Step 7 Report and/or document surface conditions as required by company policy.

Step 8 Use a **probing bar** to create holes in the ground surface along the pipeline or within the inspection area.

Step 9 Insert a gas detector drawtube in the probed holes. Do this especially down gradient of the pipeline, in gullies and low-lying areas. Examine utility system manholes for indications of hydrocarbons.

Step 10 Excavate the pipeline.

> **NOTE**
> Once the pipeline is exposed, proceed to the operator qualification requirements for Covered Task 5, *Inspect Buried Pipe When Exposed.*

9.2.0 Monitoring for Pressure Loss

Pipeline maintenance personnel must be able to monitor instrumentation for indications of a pressure loss during or after extensive earthmoving operations, such as excavation, blasting, boring, or tunneling, that are performed in the vicinity of the pipeline. Follow these steps to monitor for pressure loss:

Step 1 Identify the pipeline and segment to be monitored upon notification that an extensive earthmoving operation is taking place in the field.

Step 2 Determine normal trend pressure variance.

> **NOTE**
> Trending/pressure alarming requirements vary from line to line and vary if a line is active or inactive. There are, however, general ranges of pressure that appear normal or reasonable based on the hydraulics involved.

Step 3 Investigate any low-pressure trend outside of normal conditions, and review **historical trend** data. Immediate response to violations of pressure trending parameters will minimize impact.

> **WARNING!**
> While performing a covered task, you may encounter a pressure that is different from the expected pressure. The pressure may be too high, too low, or even absent. The change must be immediately investigated. If the cause is not determined, shut down the operating lines according to the AOC shutdown procedure, and proceed with emergency response notifications.

Step 4 Determine why the pressure violated the normal trend pressure variance.

> **NOTE**
> If the cause can be identified, go to Step 5. If not, go to Step 6.

Step 5 Communicate the violation and its cause to the appropriate persons.

Step 6 Shut down the line, referring to abnormal condition shutdown procedures.

Step 7 Communicate the violation to the appropriate persons for follow-up investigation/response. It is necessary to initiate investigation or activate emergency response from operation, engineering, and field departments per company policy.

10.0.0 ◆ CONDUCTING A LEAKAGE SURVEY

A leakage survey is performed by means of a visual inspection or by use of leak detection instruments that detect the presence of natural gas, propane, or other gases. If a leak does exist, the inspector must be alert to an unexpected gas release or the possibility of fire or explosion.

Preparation for the task includes being familiar with company procedures for emergencies, wearing appropriate personal protective equipment, and understanding the company criteria for classifying leaks. It is also necessary to understand the characteristics of the gas in the pipeline. An important characteristic is the **explosive limits**, or the range that will support combustion of a flammable gas. Each gas has its own range, typically 5% to 15% gas in air. The **lower explosive limit (LEL)** is the lowest percentage of gas in air that will support combustion or explosion. The **upper explosive limit (UEL)** is the highest concentration of a gas that will support combustion.

Elements of conducting a leakage survey include conducting a **vegetation survey**, using a **combustible gas detector**, and using a **flame ionization gas detector**.

10.1.0 Conducting a Vegetation Survey

Vegetation surveys are performed as a part of routine ROW inspections and as part of the job whenever maintenance personnel are working on the pipeline route. It is often the first alert to problems. Vegetation and other surface conditions will be affected by the escape of gas or other hydrocarbon into the soil surrounding the pipeline. A vegetation survey may include the following steps:

Step 1 Identify the portion of the pipeline system to be surveyed to ensure that the intended system is surveyed.

Step 2 Ensure that you understand the growth characteristics of the local plants so that you can correctly identify a plant problem that is caused by a natural gas leak.

Step 3 Locate the center line of the pipeline being surveyed to ensure that the area immediately above the pipe is surveyed.

> **NOTE**
> If an instrument is required to pinpoint the source of the leak, proceed to Step 7. If not, proceed to Step 8.

Step 4 Turn on and calibrate the pipe locator to ensure that it is working properly.

Step 5 Locate the pipe to be surveyed.

Step 6 Visually survey the surface conditions over the pipeline, looking for signs of abnormal changes in vegetation growth to identify possible points of natural gas leakage.

Step 7 Turn on and calibrate the instrument being used.

Step 8 Pinpoint the source of the leak.

Step 9 Based on the percent gas, volume, and location, classify the leak by utilizing the company leak-grading criteria. This helps to determine the severity of the leak.

> **WARNING!**
> Depending on the location and size of the leak, a hazardous condition may exist that could result in an explosion and/or fire. Determine the extent of the leak and act accordingly. Appropriate actions may include evacuating affected buildings or the immediate area, calling local emergency services, and contacting the company emergency center.

Step 10 If the leak classification is serious, follow company procedures to respond to serious leaks that may pose a risk to people, the environment, and property.

10.2.0 Surveying with a Combustible Gas Detector

A combustible gas detector is an instrument that is capable of detecting hydrocarbons at very slight concentrations. This instrument pulls a volume of the atmosphere being sampled through itself to analyze for the presence of hydrocarbons. The sample may be obtained manually or by automatic means. Follow these steps to use a combustible gas detector:

Step 1 Identify the portion of the pipeline system to be surveyed.

Step 2 Locate the center line of the pipe being surveyed to ensure that the area immediately above the pipe is surveyed. Use a pipe locator if necessary.

> **NOTE**
> If it is necessary to use a combustible gas detector to perform the survey, proceed to Step 3. If not, proceed to Step 4.

Step 3 Turn on the properly calibrated combustible gas detector.

Step 4 Survey in available openings along the ROW to check locations where leaking gases might migrate to, or places where gases might accumulate.

Step 5 Pinpoint the leak sources if leak indications are found.

Step 6 Following company criteria, classify the grade of the leak based on percent gas, volume, and location. This determines the severity of the leak.

Follow company procedures if the gas leak is serious and needs immediate action in order to protect people, the environment, and property from harm.

10.3.0 Surveying with a Flame Ionization Gas Detector

A flame ionization gas detector is capable of detecting leaking natural gas or other flammable gases. The survey may be performed by walking the pipeline route with the unit or by driving over the route with the unit mounted on the vehicle. This instrument is more sensitive than any other detector and may read down to one part per million. The atmosphere is pulled into the unit by a pump. This instrument is more sensitive and responds more quickly than other instruments, increasing its ability to detect leaks at a normal walking speed.

Follow these steps to use a flame ionization detector:

Step 1 Identify the portion of the pipeline system to be surveyed.

Step 2 Locate the center line of the pipe being surveyed to ensure that the area immediately above the pipe is surveyed. Use a pipe locator if necessary.

> **NOTE**
> If a flame ionization unit is required to perform this survey, proceed to Step 3. If not, this is not an instrumented leak survey.

Step 3 Turn on the properly calibrated flame ionization detector.

> **NOTE**
> If the survey is a walking survey, proceed to Step 4. If it is a mobile survey, proceed to Step 5.

Step 4 Survey in available openings along the ROW to check locations where leaking gases might migrate to, or places where gases might accumulate.

Step 5 When performing a mobile survey, drive the survey vehicle at the speed recommended by company procedures to ensure that the flame ionization unit operates properly.

Step 6 Pinpoint leak sources if leak indications are found.

Step 7 Following company criteria, classify the grade of the leak based on percent gas, volume, and location to determine the severity of the leak.

Step 8 Follow company procedures if the gas leak is serious and needs immediate action in order to protect people, the environment, and property from harm.

11.0.0 ◆ TERMS AND DEFINITIONS

The following list highlights terms that are commonly used in pipeline maintenance work. It is important to be familiar with these terms and their definitions:

- *Bar holes* – Holes punched in the ground by use of a probe bar to allow the sampling hose and probe of a combustible gas detector to be inserted to detect gas.
- *Percent scale* – A combustible gas detector scale that reads in percent of gas in air.
- *LEL scale* – A combustible gas detector scale that reads in percent of the lower explosive level of gas in air.
- *Soap test* – A leakage test in which a soapy solution in applied to an exposed pipeline component to locate leakage. Bubbles, which are caused by the leak, will be produced.

Summary

Modern pipeline maintenance focuses on the prevention of damage and deterioration so that emergency leaks and repairs can be reduced. Damage and deterioration were often considered normal wear and tear in the past, but with improvements in materials, construction, equipment, and operating and maintenance procedures, this has changed. Many of the causes of pipe failure have been eliminated or greatly reduced.

Precise location of underground structures before any excavation begins is a critical part of the process. Not only will digging in the wrong place be a waste of time and expense, but it may also inadvertently damage the pipe and/or other structures. Location of the company's pipeline should be clearly shown on company alignment maps that are to be used along with visible markers. However, the company may not be the only pipeline or utility with structures in the area, so any plan to excavate must be preceded with alerting the One-Call system.

Maintaining the pipeline route is necessary in order to access the pipeline for inspection, maintenance, and repairs. Where the pipeline and its equipment are aboveground, they must also be maintained. Markers on the route are required by regulations to inform the public and possible excavators of the existence of the pipeline. They are also used by aerial and other patrols in locating the pipeline accurately.

Sometimes it becomes necessary to excavate pipeline sections. Any time excavation is performed, procedures are followed to ensure that the pipeline is not damaged and that other underground structures and utilities are not disturbed or damaged.

A variety of tasks may be required in order to perform pipe maintenance. Some of these may be necessary because of observed problems; some may be required, such as periodic testing; and others are necessary because of changes in the system.

Maintaining the interior of the pipeline is closely related to pipeline operation and could be considered part of it. Whether the responsibility lies with operations or maintenance, the interior must be kept clear, clean, and corrosion-free.

Inside pipe surfaces require periodic cleaning to remove deposits of wax, rust, scale, or other materials. Removal may be to restore throughput capacity, or it may be preventive maintenance to control corrosion. A major component of inspecting and maintaining the pipeline interior is the pig.

Pipeline maintenance also includes surveying for leaks after blasting operations have taken place. A leak survey includes visually inspecting the pipeline ROW directly affected by the blasting, surveying additional segments of the pipeline, and checking for leaks indicative of damage to the pipeline.

Review Questions

1. Current pipeline maintenance procedures focus on _____.

   a. repair
   b. emergencies
   c. prevention of damage
   d. welding

2. The most common method for locating underground lines is _____.

   a. the One-Call system
   b. maps
   c. strip sheets
   d. line locators

3. Overhead crossings are used when _____ crossings are not practical.

   a. tank
   b. transfer
   c. relay
   d. underground

4. Permanent markers with specific information are required by _____.

   a. ANSI
   b. DOT
   c. each state
   d. AWS

5. Buried line markers must have lettering at least _____ inch(es) high.
   a. 1
   b. 2
   c. 3
   d. 4

6. When signs are used to identify navigable waterway crossings, the lettering must be at least _____ inches high.
   a. 4
   b. 6
   c. 8
   d. 12

7. Around waterways, the use of _____ is restricted.
   a. tractors
   b. herbicides
   c. manual cutters
   d. signs

8. All excavated soil must be placed at least _____ feet away from the ditch.
   a. 2
   b. 3
   c. 4
   d. 5

9. When using a trench shield, the soil must be kept at least _____ inches below the top of the box.
   a. 10
   b. 12
   c. 16
   d. 18

10. One big advantage of epoxy resin is that it is _____.
    a. thick
    b. thin
    c. heavy
    d. asphalt based

11. The type of pipe wrapping that protects pipe from ultraviolet light is _____.
    a. polyurethane
    b. polyethylene
    c. kraft paper
    d. polyolefin

12. When tightening flanges, the studs should be tightened in a numbered sequence to approximately _____ percent of the torque value for the stud size.
    a. 15
    b. 30
    c. 45
    d. 60

13. Pigs are also called _____.
    a. go-devils
    b. cleaners
    c. brushes
    d. blades

14. The actual running of the pig is accomplished by the _____.
    a. pigging vender
    b. company personnel
    c. pigging vendor and company personnel
    d. refinery personnel

15. Pipeline maintenance personnel must be able to monitor instrumentation for indications of _____ during or after extensive earthmoving operations.
    a. coating failure
    b. temperature change
    c. misalignment
    d. pressure loss

GLOSSARY

Trade Terms Introduced in This Module

Combustible gas detector: An instrument that is capable of detecting hydrocarbons at very slight concentrations. The response time is generally 7 to 10 seconds. This instrument pulls a volume of the atmosphere being sampled through itself to analyze for hydrocarbons. The sample may be obtained by manual or automatic means.

Daylight: To expose pipe during pipe locating before excavation of the pipe.

Explosive limits: The range that will support combustion of a flammable gas. Each gas has its unique range, but typical natural gas limits are 5% to 15% gas in air. Explosive limits are also called flammable limits.

Flame ionization gas detector: An instrument more sensitive than any other combustible gas detector. It may read down to one part per million and generally responds within two seconds. The atmosphere being sampled is pulled into the unit by a pump. These units may be mounted on vehicles to allow for a mobile survey of the pipeline.

Gas detection equipment: Instrumentation used to detect the presence of combustible gases in the air, the oxygen concentration of the air, and the presence of toxins, such as hydrogen sulfide.

Historical trend: The general tendency or expected drift in pressure of a pipeline segment based on past recorded data.

Lower explosive limit (LEL): The lowest percentage of gas in air that will support combustion or explosion. Mixtures below this limit are considered too lean to support combustion.

Pipe locator: A device used in leakage surveys to locate affected pipe. It sends out radio signals from underground and traces the signal down the length of the pipe. It will work on plastic or steel pipe because the plastic pipe has a metal wire running its entire length.

Probe rod: A tee-handled steel rod used to probe for the location of the pipeline.

Probing bar: A device used to bore or probe a cylindrical hole in the earth's surface that is large enough to insert the suction probe of gas detection equipment.

Seismographic vibration readings: Recordings made by a seismograph, which measures the actual movement of the ground, making a visual record of an earth tremor or vibration. A seismograph measures the velocity of ground motion, which is interpreted into a peak particle velocity for which a limitation is set to avoid damage to the pipeline.

Upper explosive limit (UEL): The highest concentration of gas in air that will support combustion. Readings above this limit are considered to be too rich in mixture to support combustion.

Vapor cloud: A visual cloud that may occur when an evaporating liquid is mixed with the humidity of the surrounding atmosphere. An example is the cloud formed during the vaporization of liquefied petroleum gas.

Vegetation survey: Observing the growth of normal vegetation over the pipeline in order to detect any changes in the vegetation that may indicate a natural gas leak.

Vent: A device installed on one end of the portion of a pipeline that crosses under a road. The vent marks the boundary of the road right-of-way, provides an exit for fluids should the line develop a leak, and aids in locating the line.

CONTREN™ LEARNING SERIES — USER UPDATES

The NCCER makes every effort to keep these textbooks up-to-date and free of technical errors. We appreciate your help in this process. If you have an idea for improving this textbook, or if you find an error, a typographical mistake, or an inaccuracy in NCCER's Contren™ textbooks, please write us, using this form or a photocopy. Be sure to include the exact module number, page number, a detailed description, and the correction, if applicable. Your input will be brought to the attention of the Technical Review Committee. Thank you for your assistance.

Instructors – If you found that additional materials were necessary in order to teach this module effectively, please let us know so that we may include them in the Equipment/Materials list in the Instructor's Guide.

Write: Curriculum Revision and Development Department
National Center for Construction Education and Research
P.O. Box 141104, Gainesville, FL 32614-1104

Fax: 352-334-0932

E-mail: curriculum@nccer.org

Craft ________________ Module Name ________________

Copyright Date ________ Module Number ________ Page Number(s) ________

Description

____

____

____

____

(Optional) Correction

____

____

____

(Optional) Your Name and Address

____

____

____

Module 62207-02

Hydrostatic Testing

(CT 41)

Introduction to the Pipeline Operator Qualification Rule, API, and NCCER

It could be said that petroleum products keep America running. Petroleum products like oil and natural gas are the principal sources of energy in the United States and are used in many manufacturing and chemical processes. Petroleum products are primarily transported by pipeline. There are more than a million miles of liquid and natural gas pipeline in this country, ranging in diameter from two inches to four feet. These pipelines bring crude oil and natural gas from production fields to processing plants and refineries—and ultimately to the consumer.

Pipeline Operator Qualification

Petroleum products are considered hazardous, so pipeline personnel must take extra precautions and follow specific procedures when working with them. To help ensure the safe handling and transportation of petroleum products, the United States Department of Transportation (DOT) mandates safe practices as federal regulations.

One such regulation, the Operator Qualification (OQ) rule, requires pipeline operators to develop a written qualification plan to evaluate their personnel and their contract personnel (see *49 CFR Part 192 Subpart N—Gas,* and *49 CFR Part 195 Subpart G—Liquid*). The evaluation hinges on the operator's ability to perform certain covered tasks and to be able to recognize and respond to abnormal operating conditions that may be encountered while performing these covered tasks.

Covered Tasks

A pipeline performance task is deemed *covered* if the activity passes the following four-part test:

1. Is it performed on a pipeline facility?
2. Is it an operations or maintenance task?
3. Is it performed as a requirement of *49 CFR Part 195 (Liquid)* or *49 CFR Part 192 (Gas)*?
4. Does it affect the operation or integrity of the pipeline facility?

If the answer to all four questions is *yes,* then the activity is considered *covered* and must figure into the evaluation process of pipeline personnel.

American Petroleum Institute

The OQ rule and its related covered tasks provided the catalyst to develop a unified pipeline training and assessment program—so that pipeline operators and owners have ready access to the assessment and training tools they need. This effort was the direct result of the vision and leadership provided by The American Petroleum Institute (API) and API members.

Efforts to unify the industry are nothing new to API. Indeed, the impetus for forming the Institute in 1919 was the need to standardize engineering specifications for drilling and production equipment. Today, the most pressing issues revolve around public perceptions and government policies toward the industry. Speaking with one voice on these issues has become as essential as having interchangeable parts in the field. API's members determine the petroleum industry's positions on public policy as well as the standards that govern its day-to-day operations worldwide.

Consortium on Operator Qualification

API assembled a focus group whose primary purpose was to identify covered tasks and provide compliance recommendations to pipeline operators.

The Consortium on Operator Qualification (COOQ) became the guiding force for the identification and development of covered tasks for the liquid pipeline industry. Their work product, API publication 1161, *Guidance Document for the Qualification of Liquid Pipeline Personnel,* formed the basis for the resulting Pipeliner Training and Assessment Program.

The covered task list created in API publication 1161 has now evolved to include gas covered tasks and is maintained by NCCER. The most current covered task list can be downloaded at www.nccer.org.

National Center for Construction Education and Research

With the guidance and support of API, its members, and the Consortium on Operator Qualification, the National Center for Construction Education and Research (NCCER) was able to design and implement the Pipeliner Training and Assessment Program.

NCCER is a not-for-profit 501(c)(3) education foundation established in 1995 by the world's largest and most progressive construction and maintenance contractors and national contractor associations. NCCER was founded to address the

severe workforce shortage facing the construction and maintenance industries and to develop a standardized training process and curricula. Today, NCCER is supported by hundreds of leading construction and maintenance companies, manufacturers, and national associations and developed more than 35 construction and maintenance titles.

NCCER also maintains a National Registry, which contains the records of trainees and instructors who have received NCCER training. All completion information for every trainee is sent to the NCCER and filed within the Registry. The Registry can then confirm training and skills for workers as they move from company to company, state to state, or even within their own company.

Standardized Curriculum (NCCER's Contren™ Learning Series)

NCCER curricula are developed using the expertise of industry subject matter experts (SMEs). The use of SMEs as an integral part of every new development or revision ensures that the curriculum is current, industry-driven, and indicative of industry practices. The development and/or revision process is facilitated from start to finish by NCCER project managers, who work closely with the SMEs and a select team of technical writer(s) on the development and presentation of the curriculum.

Pipeline: Five Training Paths

The pipeline curriculum focuses on essential skills and competencies within five skill areas:

- Corrosion Control
- Electrical and Instrumentation
- Maintenance
- Mechanical
- Operations (Field and Control Center)

SMEs from the pipeline industry have determined the skills and competencies needed based on API's covered tasks and the consensus of the group/industry. Specific company information is generally not incorporated into the curriculum unless the SMEs determine that the information is the industry standard.

Training Modules and Covered Tasks

The smallest unit of instruction in the curriculum is a *module*. Each curriculum is composed of modules (the equivalent of stand-alone chapters) that are logically sequenced into complete *levels* of instruction. Every module is task-oriented and requires the successful completion of a written (knowledge-based) examination and a performance (hands-on) test. Modules may be individually selected and/or combined for efficient task training. Modules in the pipeline curricula, to the extent possible, will align with API-developed covered tasks. While not all modules will relate to covered tasks, all covered tasks will relate to at least one module.

Benefits of Training to the Pipeline Marketplace

Applying NCCER's standardized training process allows pipeline owners and contractors in the United States to establish the benchmark for acceptable skill and competency levels within the pipeline industry. This process will assist training organizations in developing and/or improving standardized pipeline training programs that effectively match their resources and operating conditions while responding to regulatory demands and to the needs of the industry. The process will also recognize training organizations that meet the minimum standards of competence, warranting regulatory, public, and professional confidence. This in turn will foster national unity within the pipeline industry for the mutual benefit of individual technicians, owners, contractors, and ultimately pipeline-users and society.

Additional Resources

To review the Pipeliner Training and Assessment Program Implementation and Resource Guide and the latest version of pipeline covered tasks, visit NCCER's Web site at www.nccer.org, or contact:

National Center for Construction Education and Research
PO Box 141104
Gainesville, FL 32614-1104
(888) 622-3720
Web site: www.nccer.org

For more information on API, please contact:

American Petroleum Institute
1220 L Street, Northwest
Washington, D.C. 20005
(202) 682-8000
Web site: www.api.org

To order *API Pub 1161*, contact Global Engineering Documents by calling 1-800-854-7149 or visiting their Web site at www.global.his.com.

Additional sources of information on the pipeline operator qualification regulation include:

DOT – Office of Pipeline Safety
www.ops.dot.gov

Transportation Safety Institute
www.tsi.dot.gov

COURSE MAP

This course map shows all of the modules in the second level of the Pipeline Maintenance curriculum. The suggested training order begins at the bottom and proceeds up. Skill levels increase as you advance on the course map. The local Training Program Sponsor may adjust the training order.

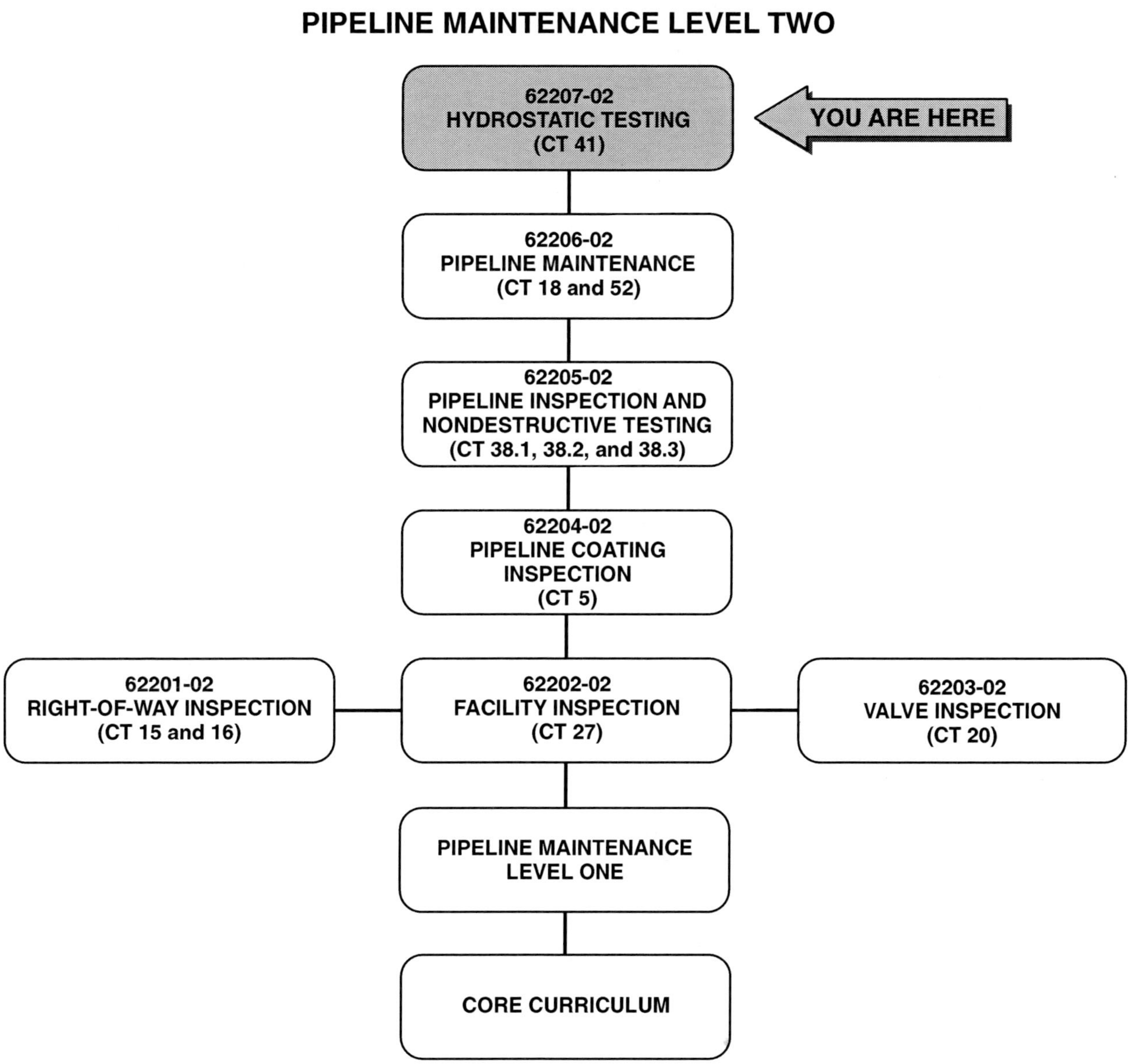

207CMAP.EPS

MODULE 62207-02 CONTENTS

Figures

Hydrostatic Testing (CT 41)

Objectives

When you have completed this module, you will be able to do the following:

1. Explain the purpose of hydrostatic testing.
2. Prepare for hydrostatic testing.
3. Perform hydrostatic testing procedures (CT 41).

Prerequisites

Before you begin this module, it is recommended that you successfully complete the following: Core Curriculum; Pipeline Core, Modules 66101-02 and 66102-02; Pipeline Maintenance Level One; Pipeline Maintenance Level Two, Modules 62201-02 through 62206-02.

Required Trainee Materials

1. Pencil and paper
2. Appropriate personal protective equipment

1.0.0 ◆ INTRODUCTION

Although **pressure testing** is not done as a part of routine inspections, most pipeline owners check the integrity of their pipelines. The most common form of pressure testing is **hydrostatic testing**. Usually done during periods of slack demand, this type of test requires that the line be taken out of service and cut at specified intervals.

Hydrostatic testing is also required by federal regulations in specific circumstances. These circumstances include new piping installations, piping that has welding alterations, overwater piping, and times specified by the piping inspector or engineer. When conducted, tests are performed in accordance with Department of Transportation (DOT) regulations.

This module covers the following:

- Hydrostatic testing
- Preparation for testing
- Test procedures

2.0.0 ◆ HYDROSTATIC TESTING

In pressure testing, the internal pressure is increased to a level above the normal or maximum operating pressure (MOP) in a segment of the pipeline for a specified length of time. The pipeline segment is isolated from the pipeline. A liquid medium, usually water, is used to fill the line and provide the internal pressure.

The pressure test is used to ensure that the piping will withstand operating pressures without leaking. If a leak is discovered during testing, the segment has failed the test and must be repaired and retested.

2.1.0 Required Testing

Testing is required for new construction and the replacement of existing pipelines in accordance with DOT regulations and other applicable regulations. Existing pipelines that need to be qualified for a higher operating pressure or that need to be requalified for the current operating pressure are required to be tested according to the same regulations.

2.2.0 Testing Process

During the testing process, manifolds are welded on to the pipe ends, and the test sections are filled with clean water. Once the manifold is welded to the section of pipe, a pig is inserted to push out air ahead of the water and to restrict the water to the test section.

A low-**head** water pump is set near the water source and connected to the test manifold by fill pipe. Several more fill pumps are often required to increase volume and to boost pressure. Filters clean the water before it enters the pipe.

When the line is filled with water, the fill pipe is disconnected, and a pressure pump is hooked up to the manifold. This pump increases the pressure of the water in the line to the specified pressure. The line is then pressure tested. Instruments called deadweight testers and electronic pressure gauges are used to monitor the pressure during the test. Temperature recorders are also used because ambient and external temperatures have an effect on pressure. Some companies use certified electronic recorders. All instruments used in the testing must be certified within one year of the test.

The required pressure must be maintained for a specified period of time for the test to be successful. When the test is complete, valves on the test header are opened, reducing the pressure. If no problems are encountered, the line is dewatered, dried, and returned to service. Compressed air, gas, or product is used to propel the pig back through the line to remove the water. Several pig runs may be needed to dry out the pipe. Other remedial work may be done on the pipeline sections while they are out of service and accessible. The test section is rejoined to the pipeline and returned to service.

3.0.0 ◆ PREPARATION FOR TESTING

Before testing can begin, a variety of tasks must be performed to prepare for the test. The first requirement is to understand the company policy and procedures involved, including those related to safety procedures and appropriate personal protective equipment. Other preparations include identifying the medium, selecting the required equipment, and planning the test.

3.1.0 Identifying the Medium

The medium used for the test should be, and normally is, clean water. However, there are some circumstances that may require other medium specifications. These circumstances may involve the particular type of metal in the piping, climate conditions, or disposal issues. In addition to disposal of the medium, the possibility of release in case of a leak must be considered. If there is the possibility of the water freezing, if the water may have adverse effects on the piping, or if the test water may become contaminated and present disposal problems, another suitable nontoxic liquid may be used.

A liquid petroleum may be used for testing under very specific circumstances. Basically, the petroleum cannot be used if the segment is offshore, near environmentally sensitive areas, or near cities or highly populated areas. During testing, nearby facilities must be unoccupied, the segment must be kept under surveillance, and suitable response and cleanup procedures must be in place.

3.2.0 Planning the Test

Once the need for the test is determined, a good deal of planning must be done to prepare for it. Planning includes becoming familiar with the past and future operating pressures of the particular pipeline, becoming familiar with the compositions of the pipe and piping components, becoming familiar with the ratings of all pipeline valves, equipment, and other connections, and anticipating the temperatures involved. It also involves dealing with safety and environmental requirements.

3.2.1 Pressure/Temperature Plans

The pressure of the test should not approach the **specified minimum yield strength (SMYS)** of the material in the pipeline unless special care is taken to prevent the pipe from overstraining. Documents on the grade and wall thickness of the pipe and any internal inspection data should be examined to ensure the integrity of the pipe for the expected pressures. A pressure volume plot during pressurization may be needed if the tests expect levels near SMYS.

Comparisons of the ratings for pipeline components must be made to determine which component has the lowest rating; thus, the maximum possible test pressure is controlled. If valve ratings are to be exceeded, the valves may be removed or blanked for the test or held down by safety clamps. Other fittings, such as gauge glasses, pressure gauges, expansion joints, and rupture disks, may need to be removed or blanked.

Anticipated temperatures have a direct effect on the pressure, so plans must consider the temperature of the medium, the atmosphere, and the ground. Large amounts of exposed piping can result in large temperature-related pressure changes, making a stable test difficult. Plans must also be made for preventing the freezing of exposed pipe during cold weather.

3.2.2 Safety/Environmental Plans

The medium and other treatments that are to be used for the test need to be handled and disposed of according to state and local regulations. These must be reviewed to determine what procedures will be used. There must be a sampling procedure in place to check the medium before and after the test. Locations and means for disposal of the medium must be selected.

Documentation for the pipeline to be tested must be gathered and reviewed to determine the exact location, depth, facilities, and components of the segment. Profile and alignment maps and other drawings can help with this. The proper authorities, including landowners, must be notified of the planned testing.

Personnel to be involved in the testing should be selected and assigned specific responsibilities. Safety precautions and procedures for testing personnel must be reviewed. Precautions and procedures for protecting the public must also be reviewed and put into place. A plan for dealing with problems must be drawn up in case of pipe test failure, equipment failure, personnel mistakes, and/or emergency situations.

3.2.3 Test Procedure Plan

A plan of the actual test procedure must be drawn before the actual test. This plan serves as a guide through the test and ensures that correct procedures are followed throughout. It may also help to point out possible problems to be solved.

The plan should show relevant notes, data, and details of the lengths, locations, and elevations of the test segment. Locations of taps for the pressure-recording devices must be indicated. The medium must be described along with the volume needed to fill and pressurize the pipe and the fill rates expected. The minimum test duration must be indicated, as well as an identification of the weakest link or component that controls the maximum pressure. Other information in the plan includes the methods to be used for:

- Cleaning, decommissioning, filling, and recommissioning the test segments
- Pressurizing the test segments
- Isolating the test segments
- Removing and disposing of the medium
- Providing safety precautions

3.3.0 Selecting Required Equipment

All of the equipment that is needed must be checked for proper operation and gathered at the test site. Measuring equipment that is used must be rated for the expected pressures, calibrated, and certified. Most tests use this method, but always follow company policies. The equipment that is required may include:

- Pumps
- Manifolds
- Valves
- Tanks
- Measuring instruments
- Miscellaneous equipment

3.3.1 Pumps

Three main pumps are used during the pressure test. A high-volume pump is used to fill the line and ensure adequate pressure to overcome the head. This pump must also maintain sufficient velocity to move debris, minimize interfaces, ensure turbulent flow, and keep pigs moving.

An injection pump is used to introduce corrosion inhibitors, leak-detection dyes or gases, or other chemicals into the test segment if their use is required.

A variable-speed positive-displacement pump pressurizes the line and must be capable of pressurizing to a level that exceeds the desired test pressure. The pump must have a known volume per stroke and must be equipped with a stroke counter. A constant-speed pump with a variable flow rate may be used instead if the volume of the medium can be measured during pressurization.

3.3.2 Manifolds

The manifolds and temporary connections are to be fitted onto both ends of the pipeline segment. The manifold consists of a section of pipe with a weld cap at one end, a pipe riser for filling purposes, and several small connections for pressurizing and taking readings. These manifolds and connections must withstand the expected pressures during testing.

3.3.3 Valves

A manual valve is used to relieve pressure to prevent overpressuring of the test segment during pressurization and the test. If a test is to be maintained for a period of time and there is the possibility of thermal expansion, overpressure may occur.

3.3.4 Tanks

A portable tank or transport into which excess test medium can be discharged and from which make-up volumes can be drawn may be needed. Tanks or transport for receiving the test medium after the test may also be required if on-site disposal is not possible.

3.3.5 Measuring Instruments

The instruments providing measurements during the test are the largest category of equipment required. All instruments must be calibrated according to company policy. These include:

- A pressure sensing and display device that has the pressure range and increment divisions needed
- A deadweight tester or other pressure-sensing device capable of measuring in increments of 1 pound per square inch or less
- A continuously recording pressure-measurement device, like a chart recorder, that provides a permanent record of the pressure over time
- A test medium temperature-sensing and display instrument suitable for expected test temperatures and capable of reading in increments of no less than 0.1°F
- A continuously reading temperature-measurement device that provides a permanent record of pipeline temperature versus time
- Equipment that measures ambient temperature
- Electronic pressure/temperature monitoring and recording systems that assist in the analysis of the test data
- A device to measure any excess test medium that is removed

3.3.6 Miscellaneous Equipment

Other equipment needed for the test include pigs, scrapers, and other devices to clean and dry the line. Weather protection, such as insulation, may be needed to protect instruments and communications equipment from excessive cold or heat. Replacement and repair items in case of pipe failure and filters for the water source must be available. Equipment used to measure how much of the medium is drawn off as it is needed is also used during the test.

4.0.0 ◆ TEST PROCEDURE

When all the preparations are made, the piping to be tested is taken out of service and emptied of product. The actual test procedure begins with isolating the segment and connecting the manifolds at each end. The taps for recording devices are located at the end of the tested section and according to company policy. This section discusses:

- Cleaning the line
- Filling the line
- Pressurizing the line
- The test period
- The test records
- Displacing the medium

4.1.0 Cleaning the Line

Although pigging has been performed to empty hydrocarbons out of the line, this does not normally remove all of the product. Cleaning is necessary to remove as much of the product as possible. A train of cleaning pigs is run to remove sediments, paraffins, and other residue. It is also advisable to run a sizing pig or caliper tool to identify any geometric abnormalities that may exist.

4.2.0 Filling the Line

The filling operation is performed with the large-volume fill pump, using filter screens if a non-municipal water supply is being used, a flowmeter to measure flow rate, and filling pigs. A portable tank may need to be available for making up the differences between the actual water supply and the pumps.

The fill pump must provide a sufficient volume so that the fill pigs travel at a speed that maintains a tight fit for the pigs and reduces the chance of introducing air behind them. Care must be taken so that the water column does not cause the pigs to travel faster than the fill rate, which avoids getting air or gas behind the pigs. This could happen on a slope if there is no form of backpressure.

The flowmeter is monitored as the pipe is filled so that adjustments can be made as necessary. As the pressure builds, the flow rates drop. The flowmeter also helps to match the actual volume with the calculated fill volume. The temperature of the test medium must be taken prior to putting it into the line to help determine line stability.

4.3.0 Pressurizing the Line

The pressurization of the line begins once the line is full of fluid and the appropriate measures have been taken to bleed any and all air or gas out of the section. The line is then brought up from static pressure to the desired test pressure that was determined beforehand. Pressurization must occur at a controlled rate to avoid surging the line. This should always follow company policy. The following equipment is used for this process:

- High-pressure positive-displacement pump
- Flowmeter and volume accumulator
- Deadweight pressure tester
- Temperature recorder
- Pressure recorder
- Portable tank

During the pressurization process, always check connections periodically for leaks, monitor and log the flow rate, and check the actual volume of fluid versus the calculated volume to pressurize.

WARNING!
If unexpected hazardous liquids or gases are discovered or if hydrocarbons trapped in the system during the test are released due to component failure, local emergency services should be notified immediately as should the company.

WARNING!
In case of the ignition of hazardous liquids or gases resulting in a fire or explosion, the area must be evacuated and local emergency services and the company must be contacted immediately.

The rate is reduced once the pressure reaches 80% to 90% of the desired test pressure, especially when the test is at or near 100% SMYS. Once the target pressure is reached, the equipment is stopped and isolated from the piping segment. All valves and connections must be checked for leakage. A stabilization period at this time allows the temperature and pressure to settle before the test period.

WARNING!
The high pressure maintained in the pipe is dangerous. Test personnel must stay a safe distance from the test section during the test period.

4.4.0 Test Period

During the stabilization period, instruments must be checked to ensure that the test pressure is being maintained and temperatures have stabilized. When the system has stabilized, the test period timing should begin.

Pressure and temperature must be continuously monitored and recorded. Deadweight tester comparisons with pressure recorder readings should be made at the beginning of, at the end of, and periodically during the test. Any weather changes must be recorded because they could affect the recording charts.

Minor or gradual pressure changes during the test may be a result of air in the segment, temperature, or leaks through small defects or loose flange connections. Extending the test can demonstrate that air and temperature changes have been accounted for.

CAUTION
Unexplained pressure deviations might be the result of pipe or component failure during the test or the presence of air in the test medium. The cause of the deviations should be determined, and the appropriate steps to correct the deviations should be taken. Additionally, the company should be notified as necessary.

CAUTION
In case of the failure of components such as the pipe, valves, gaskets and threaded fittings, the damage should be assessed, and the appropriate company personnel should be notified.

According to *DOT 195.304*, the test pressure for each pressure test conducted must be maintained throughout the segment of the system being tested for at least 4 continuous hours at a pressure equal to at least 125% of the maximum operating pressure. In the case of a pipeline that is not visually inspected for leakage during the test, such as an underground pipeline, the pressure must be maintained for at least an additional 4 continuous hours at a pressure equal to at least 110% of the maximum operating pressure *(DOT 192.505)*.

4.5.0 Test Records

Test records that are necessary include field records and required file records. A complete record of the pressure test is required, including such details as weather variations, test failures, bleedoffs, and repressurizations. If any failures occur, they must be described on the records with details such as the exact location, type, and cause. The method of repair must also be described. Components that fail and are replaced must be marked with their location and with the pressure that caused them to fail.

Certain records are required by federal regulations to be kept on file by the operator, including:

- The continuous pressure vs. time record (*Figure 1*)
- The temperature vs. time record (*Figure 2*)
- The deadweight calibration certificate
- The test plan

- The pressure test record and certification (*Figure 3*), including:
  - Qualification calculations (*Figure 4*)
  - Pressure and temperature (*Figure 5*)
  - Record of failures (*Figure 6*)
  - Profile of the pipeline test section
  - Drawing of any facility piping tested
- The name and signature of the operator

These records are kept for the life of the facility or until new test records are generated.

4.5.1 Test Records for Liquid Pipelines

A record must be made of each pressure test that is required, and the record of the latest test must be retained as long as the facility tested is in use. The records required must include:

- The pressure recording charts
- The test instrument calibration data
- The name of the operator
- The name of the person responsible for performing the test
- The name of the test company used, if any
- The date and time of the test
- The minimum test pressure
- The test medium
- A description of the facility tested and the test apparatus
- An explanation of any pressure discontinuities
- The points where elevation differences in the section under test exceed 100 feet

4.5.2 Test Records for Gas Pipelines

Each operator shall make, and retain for the useful life of the pipeline, a record of each test performed. The record must contain the following information:

- Operator's name
- Name of the operator's employee responsible for performing the test
- Name of any test company used
- Test medium used
- Test pressure
- Test duration
- Pressure recording charts
- Elevation variations
- Leaks and failures noted and their dispositions

4.6.0 Displacing the Medium

The test water may be displaced with liquid petroleum, air, or inert gas. Spheres or other pigging devices may also be used to displace the medium. The test water is disposed of in accordance with all environmental regulations even if it must be stored until disposal permits are obtained. The pipeline is usually dried after displacing the water to ensure product quality and internal corrosion control.

5.0.0 ◆ CONDUCTING A PRESSURE TEST

When conducting a pressure test, there are many aspects that must be considered. To properly conduct a pressure test, you must be able to:

- Identify the test duration and pressure required for the planned test. Ensure that the test is performed in accordance with the test plan and *DOT 195* and *192*. Additionally, ensure that the test pressure range justifies the desired MOP.
- Identify if state pipeline safety regulations for a pressure test differ from *DOT 195* and *192*. You should recognize and understand any differences between a state-mandated test and *DOT 195* and *192* test.
- Identify the test medium. You should understand the difference between testing with a liquid and testing with a gas.
- Know company policy on the test procedure.
- Explain the fill process. You should understand the proper means for filling a pipeline with the test medium.
- Explain the stabilization process. You should understand the dynamics of the changes that occur with the test medium and the pipe as the two reach a thermal equilibrium.
- Understand the method(s) to pressurize the item being tested. You should understand the different methods that may be used to pressurize the pipe to ensure a valid test.
- Explain how to factor in pressure and temperature into the test results. You should be able to factor in any change to ensure a leak-free test.
- List records that must be produced. You should know which records must be generated and maintained to validate the pressure test.

SAMPLE PRESSURE RECORD

Company ______________________
System ______________________

Description of instrument (make/model) ______________________
Serial number of instrument ______________________
Test section no. ______________________ ______________ miles
MP ____________ to MP ____________ Station no. ____________ to Station no. ____________
Location of chart recorder MP ______________________ Station no. ____________
Start: Time ______________________ Date ______________________
End: Time ______________________ Date ______________________
Contractor rep. ______________________ Title ____________ Date ____________
Pipeline company rep. ______________________ Title ____________ Date ____________
Project engineer ______________________ Date ____________

Notes:
MP = mile post
This pressure information should be included on the permanent record of pressure versus time. Placing this information on a stick-on label and sticking the label to the permanent record might be considered.

SAMPLE

207F01.EPS

Figure 1 ◆ Pressure vs. time record form.

SAMPLE TEMPERATURE RECORD

Company ______________________
System ______________________

Description of instrument (make/model) ______________________
Serial number of instrument ______________________
Test section no. ______________________ ______________ miles
MP ____________ to MP ____________ Station no. ____________ to Station no. ____________
Location of chart recorder MP ______________________ Station no. ____________
Start: Time ______________________ Date ______________________
End: Time ______________________ Date ______________________
Contractor rep. ______________________ Title ____________ Date ____________
Pipeline company rep. ______________________ Title ____________ Date ____________
Project engineer ______________________ Date ____________

Notes:
MP = mile post
This temperature information should be included on the permanent record of temperature versus time. Placing this information on a stick-on label and sticking the label to the permanent record or using a rubber stamp might be considered.

SAMPLE

207F02.EPS

Figure 2 ◆ Temperature vs. time record form.

SAMPLE PRESSURE TEST RECORD AND CERTIFICATION

Test section ______________________

Date ______________________

Company ______________________ System ______________________

Description from ______________________ to ______________________

New construction ❍

Requalification ❍ Replacement or relocation ❍ Pipeline ❍ Station ❍

Test medium: water ❍ Other ______________________ Inhibitor ______________________

Design data code: ❍ B31.4 ❍ B31.8 (Appendix N) ❍ Other ______________________

Pipe design data

Specification and grade	Weld joint factor	Design factor	OD	Wall thickness	SMYS	Design pressure	Comments

Pressure test:

Test pressure should be as follows:

__________ Minimum at high point __________ % SMY __________ Maximum at low point __________ % SMY

Elevations: Low point ______________ High point ______________ DWT ______________

Qualifications:

Date of test ______________________ Duration of test ______________________

Testing and recording witnessed by ______________________ Date ______________

Company ______________________ Title ______________

Company representative ______________________ Title ______________

Testing pressure:

Maximum at low point ______________________ for % SMYS ______________

Minimum at high point ______________________ for % SMYS ______________

Qualified to operate at ______________________ for % SMYS ______________

Report checked by ______________________ Date ______________

Approved by ______________________ Time ______________

Testing company ______________________

Attached documents:

Pressure record ❍ Pressure and temperature log ❍ Temperature record ❍

Test instrument calibration data ❍ Sketch or diagram ❍ Qualification calculations ❍

Profile ❍ Failure records ❍

Comments:

____

____

____

____

Notes:

OD = outside diameter, SMYS = specified minimum yield strength, SMY = specified minimum yield, DWT = deadweight tester

For test sections containing more than one type of pipe, the maximum test pressure at low point, minimum test pressure at high point, and resulting SMYS need to be calculated for each type of pipe.

207F03.EPS

Figure 3 ◆ Pressure test record and certification form.

SAMPLE QUALIFICATION CALCULATIONS

Company ____________________
System ____________________

Pipeline Data (at controlling location) ____________ OD ________ ________ WT Specification
Internal pressure at SMYS ____________ psi
Test medium: Fresh water? (Yes, at 0.433 psi/ft) (No) Other ________ at ________ psi/ft
Design test pressure: Maximum ____________ psi Minimum ____________ psi
Test section number ____________
From station ________ (MP ________) to station ________ (MP ________)
Time and dates: From ________ (Time) hours ________ (Date) to ________ (Time) hours ________ (Date)

Deadweight tester data
Location: Station ________ (MP ________) Elevation ________ ft
Tester pressure: Beginning ________ psi ending ________ psi minimum ________ psi
Acceptable? (Yes) (No)
Use tester pressure of ________ (P) psi at station ________ Elevation ________ (E)

Minimum Pressure in Test Section[a]	**Maximum Pressure in Test Section**[a]
Highest elevation in test section ________ (H) ft	Lowest elevation in test section ________ (L) ft
Location: Station ________ (MP ________)	Location: Station ________ (MP ________)
Difference in elevation from tester:	Difference in elevation from tester:
________ (H) – ________ (E) = ________ (H) – (E) ft	________ (E) – ________ (L) = ________ (E) – (L) ft
Pressure at highest elevation:	Pressure at highest elevation:
= Tester pressure – (Difference in elevation x psi/ft)	= Tester pressure – (Difference in elevation x psi/ft)
= ________ (P) – (________ (H) – (E) x ________)	= ________ (P) – (________ (E) – (L) x ________)
= ________ = Minimum test pressure	= ________ = Maximum test pressure
= ________ % SMYS	= ________ % SMYS

Maximum allowable operating pressure in this test section = 72% SMYS = ____________ psi or
80% minimum test pressure = ____________ psi or
or controlling component ____________ (circle whichever is lowest).
Calculated by ____________ Approved by ____________
Remarks ____________
____
____________ Date ____________

Notes: OD = outside diameter, wt = wall thickness, SMYS = specified minimum yield strength, MP= mile post, p = pressure, E = elevation, H = highest, L = lowest
[a]For test sections containing more than one type of pipe, the 72-percent SMYS minimum test pressure and resulting percentage of SMYS need to be calculated for each type of pipe and used in determining the maximum allowable operating pressure.

207F04.EPS

Figure 4 ◆ Qualification calculations form.

SAMPLE PRESSURE AND TEMPERATURE LOG

Company ______________________

System ______________________

Test section no. ______________ From station no. ______________ MP ________

to station no. ______________ MP ________

Pressure/temperature sensor station no. ______________

Start of test period: Time ______________ Date ______________

End of test period: Time ______________ Date ______________

No.	Time	Pressure (psig)	Pipe temp.	Ambient temp.	No.	Time	Pressure (psig)	Pipe temp.	Ambient temp.
1.					25.				
2.					26.				
3.					27.				
4.					28.				
5.					29.				
6.					30.				
7.					31.				
8.					32.				
9.					33.				
10.					34.				
11.					35.				
12.					36.				
13.					37.				
14.					38.				
15.					39.				
16.					40.				
17.					41.				
18.					42.				
19.					43.				
20.					44.				
21.					45.				
22.					46.				
23.					47.				
24.					48.				

SAMPLE

207F05.EPS

Figure 5 ◆ Pressure and temperature log.

SAMPLE PRESSURE TEST RECORD OF FAILURE

Company ______________

System ______________

Line section ______________

From station no. ______________ to station no. ______________

Date	Time	Leak station no.	Leak elevation	Leak (psi)	Deadweight elevation	Deadweight (psi)	Deadweight station	Failure description

207F06.EPS

Figure 6 ◆ Record of failures form.

Summary

Hydrostatic tests are used to determine if the complete pipeline is capable of withstanding, without leakage or failure, internal pressures higher than proposed operating pressures.

This module discusses the purpose of hydrostatic testing and the documentation required when recording information obtained during a test. Preparation for performing a hydrostatic test is also covered. A description of the hydrostatic testing procedures and the general testing process is included along with sample forms, records, and logs.

Review Questions

1. The most common form of pressure testing is _____ testing.

a. wall thickness
b. hydrostatic
c. pneumatic
d. vacuum

2. The liquid medium used for pressure testing is usually _____.

a. water
b. oil
c. liquid petroleum
d. liquid nitrogen

3. To fill the pipe for a pressure test, a(n) _____ water pump is used.

a. high-head
b. vacuum
c. alternating
d. low-head

4. Temperatures have a direct effect on the _____ during a test.

a. pipe
b. components
c. medium
d. pressure

5. Disposal of the medium requires that a proper _____ be selected.

a. additive
b. location
c. type
d. storage container

6. During a pressure test, _____ main pumps are used.

a. five
b. four
c. three
d. two

7. The test medium _____ must be known before putting it into the line.

a. temperature
b. pressure
c. weight
d. corrosiveness

8. The test pressure for each pressure test must be maintained for at least _____ continuous hours.

a. four
b. six
c. eight
d. ten

9. Test records are kept _____ or until new test records are generated.

a. for 5 years
b. per company policy
c. for the life of the facility
d. per state regulations

10. The pipeline is usually _____ before being put into service after a test.

a. lubricated
b. dried
c. sanded
d. X-rayed

GLOSSARY

Trade Terms Introduced in This Module

Head: (1) The height of a column of liquid required to produce a specific pressure; it is more properly referred to as hydraulic head, which is the force exerted by a column of liquid above the point at which the pressure is measured. Although head refers to distance or height, it is used to express pressure, since the force of the liquid column is directly proportional to its height. In a pipeline, it is the height of a theoretical column of fluid extending from the lowest level of a line or tank to a given point of elevation, and it is usually measured in feet. (2) On a compressor, the attachment or part of the machine used to seal the end of the cylinder barrel that contains the operative devices, such as valves.

Hydrostatic testing: The most common final quality check on a complete pipeline. Water is pumped into the line and pressured up to a specified level. The pipe is then checked for leaks or ruptures.

Pressure testing: The application of pressure to a pipeline using air, liquid, natural gas, or inert gas as the test medium.

Specified minimum yield strength (SMYS): The specified minimum yield strength of steel pipe expressed in pounds per square inch (psi), as defined in DOT regulations.

CONTREN™ LEARNING SERIES — USER UPDATES

The NCCER makes every effort to keep these textbooks up-to-date and free of technical errors. We appreciate your help in this process. If you have an idea for improving this textbook, or if you find an error, a typographical mistake, or an inaccuracy in NCCER's Contren™ textbooks, please write us, using this form or a photocopy. Be sure to include the exact module number, page number, a detailed description, and the correction, if applicable. Your input will be brought to the attention of the Technical Review Committee. Thank you for your assistance.

Instructors – If you found that additional materials were necessary in order to teach this module effectively, please let us know so that we may include them in the Equipment/Materials list in the Instructor's Guide.

Write: Curriculum Revision and Development Department
National Center for Construction Education and Research
P.O. Box 141104, Gainesville, FL 32614-1104

Fax: 352-334-0932

E-mail: curriculum@nccer.org

Craft ____________________ Module Name ____________________

Copyright Date __________ Module Number __________ Page Number(s) __________

Description

____

____

____

____

(Optional) Correction

____

____

____

(Optional) Your Name and Address

____

____

____

Index

Index